I0191901

Francis Day

Supplement to the Fishes of India

Francis Day

Supplement to the Fishes of India

ISBN/EAN: 9783744763035

Printed in Europe, USA, Canada, Australia, Japan

Cover: Foto ©berggeist007 / pixelio.de

More available books at **www.hansebooks.com**

Oardad

Division of Fis
U. S. National M

Price 5s 6d.

SUPPLEMENT

TO THE

FISHES OF INDIA;

BEING

A NATURAL HISTORY

OF

THE FISHES

KNOWN TO INHABIT THE SEAS AND FRESH WATERS

OF

INDIA, BURMA, AND CEYLON.

BY

FRANCIS DAY, C.I.E., F.L.S., & F.Z.S.,

Knight of the Crown of Italy, Hon. Member Deutscher Fischerei-Verein, and of the American Fisheries Society, President Cheltenham Natural Science Society, Vice-President of the Cotswold Naturalists' Field Club, etc., Deputy Surgeon-General Madras Army (retired), and formerly Inspector-General of Fisheries in India.

Published by
WILLIAMS AND NORGATE,
14, HENRIETTA STREET, COVENT GARDEN, LONDON
AND 20, SOUTH FREDERICK STREET, EDINBURGH.

SUPPLEMENT

TO THE

FISHES OF INDIA;

BEING

A NATURAL HISTORY

OF

THE FISHES

KNOWN TO INHABIT THE SEAS AND FRESH WATERS

OF

INDIA, BURMA, AND CEYLON.

BY

FRANCIS DAY, C.I.E., F.L.S., & F.Z.S.,

Knight of the Crown of Italy, Hon. Member Deutscher Fischerei-Verein, and of the American Fisheries Society, President Cheltenham Natural Science Society, Vice-President of the Cotswold Naturalists' Field Club, etc., Deputy Surgeon-General Madras Army (retired), and formerly Inspector-General of Fisheries in India.

Published by

WILLIAMS AND NORGATE,

14, HENRIETTA STREET, COVENT GARDEN, LONDON;

AND 20, SOUTH FREDERICK STREET, EDINBURGH.

1888.

FISHES OF INDIA.

SUPPLEMENT, OCTOBER, 1888.

During the ten years which have elapsed since the publication of my "Fishes of India," many new piscine forms have been obtained from both the seas and fresh waters of that portion of Asia. Extended investigations among specimens in European museums have likewise shown me that some species which I formerly considered as undescribed, had no title to that designation, while several of my new species have been redescribed as novelties in the Proceedings of Societies, in scientific journals or other publications. I am also indebted to Mr. Thurston of the Madras Museum, and Mr. Haly of the Colombo Museum, for some new Indian forms. I have likewise had the opportunity of consulting the volume of beautiful coloured drawings of Burmese fishes with their descriptions by the late Col. Tickell, several of which I have included in the present supplement.

As it is now proposed to re-issue my work in a more portable size, one better suited for travellers and collectors, and in which it would be very inconvenient, on account of the necessary space it would require, to embrace the entire synonymy of every species, I have considered it advisable to complete the original work to the present date. By this means the references would be available for those who are concerned in working out each species, for doubtless a vast number of Indian fishes still remain to be discovered.

Page 9. Lates calcarifer. Add synonym.

Pseudolates carifrons. Alleyn and Macleay, Pro. Lin. Soc. N. S. W. i, p. 262, pl. iii.

Page 9. Cromileptes altivelis. Add to synonyms as varieties.

Serranus striolatus, Günther and Playfair, Fish. Zanzibar, p. 11, pl. iii, f. 2.
        „   *gibbosus,* Boulenger, Proc. Zool. Soc. 1887, page 654.

The chief structural reason why these two varieties have been separated from the original form appears to be in consequence of the comparatively low spinous dorsal fin: and the colours in which last *S. striolatus* and *S. gibbosus* likewise differ. *S. striolatus* has its third and fourth dorsal spines, said to be the longest (both are broken in the single example) and one-third the length of the head (thus differing from *S. altivelis*). In the figure, the third spine is one-fourth longer than the last, which, however, exceeds the penultimate one (which is also broken), while it is more than half the height of the longest dorsal ray (1·4 to 2·4). The height of the soft portion of the dorsal and anal fins equals that of *altivelis*, but the pectoral is shorter. In colours it has fifteen or more rows of short brown streaks and numerous brown spots on the head and body. The single skin is 19 inches in length.

Serranus gibbosus. The unique specimen has been preserved in spirit, and is 15 inches in length. The third dorsal spine is longer than the last, which is one-third shorter than the longest ray. Its colours are an advance from *striolatus* towards *altivelis*, it being generally brown with spots or blotches of a larger size than in *striolatus*, some of which are surrounded by a light ring, but having a tendency to form about seventeen lines along the side.

I think these two new species cannot be specifically separated, but may be varieties of *altivelis*, which latter has not been recorded from the Red Sea, nor known to extend to Muscat or Zanzibar. If, therefore, they are varieties, they are evidently outlying specimens which show a variation in the length of the dorsal spines from what obtains in those captured more to the east. The pectoral fin appears rather short in *striolatus*, but more of the normal length in *gibbosus*, whereas in all the number of scales appears to be the same.

5 н

The figure of *Serranus altivelis* in Cuv. and Val. ii, pl. xxxv, shows the spines of the dorsal fin increasing in length to the last which is delineated nearly twice as long as the second. Cantor, "Malayan Fishes," found these spines from the third to be of nearly equal length. Bleeker shows them slightly, but gradually increasing to the last, which is figured one-fifth longer than the third. I have observed them more corresponding with Cantor's description, but all have been small specimens. Still the foregoing show that differences do exist in the length of these spines, that augmenting from the third to the last is not universally carried out in the same way. In *altivelis* the length of the third dorsal spine is about ¼ of that of the head, in the larger *gibbosus* 15 inches long it is 2⅙, and in the still larger *striolatus* ⅙.

Page 12.   SERRANUS AREOLATUS. Add to synonymy.
     ?   ,,   ,,   *wandersi*, Bleeker, Atl. Ich. vii, p. 47, Perc. t. iii, f. 1.
         ,,   ,,   *geoffroyi*, Klunz. Fische Rothen Meere, p. 3.
         ,,   ,,   *multipunctatus*, Koss. u. Raub. p. 6.
    Klunzinger considers *Perca areolata*, Forsk., to be identical with *Serranus angularis*, C. V.

Page 13.   SERRANUS UNDULOSUS.
    Steindachner considers that among the synonyms of this fish may be included *Serranus acutirostris*, C. and Val., *S. fuscus*, Lowe, *S. tinca*, Cantraino and *S. macrogenis*, Savi.

Page 16.   SERRANUS TUMILABRIS. Omit from synonyms.
         ,,   ,,   *summana* Cuv. and Val.: Rüppel : and Lefèb.
         ,,   ,,   *tumilabris*, Cuv. and Val.: Günther Catalog.
    Add *Anthias argus*, Bloch, t. cccvii.

Page 17.   SERRANUS DIACANTHUS. Add synonym.
    Epinephelus retouti, Bleeker, Fish. Madagascar, p. 21, pl. xii, f. 1.

Page 19.   For SERRANUS MALABARICUS read S. PANTHERINUS. Add synonym.
    ? *Holocentrus malabaricus*, Bl. Schn.

Page 21.   SERRANUS MORRHUA. Add to synonymy.
    Serranus præopercularis, Boulenger, P. Z. S. 1887, p. 654.

    As I have remarked, and likewise figured, the young of this species has broad white bands, whereas, as it becomes older, it is brown with narrow black lines, which were the original borders of the white bands. In the Paris Museum is a young specimen in which there are dark spots along these lines, while among these percoid fishes longitudinal dark bands or lines have a tendency to become spots, and finally disappear, as horizontal bands have to fade away. In Klunzinger's figure, F.R.M. t. i, f. 2, three brown bands radiate from the eye and become four curved ones on the body, the first going to the eighth dorsal spine, the second to the fifth ray, while between them are blotches, spots or markings of the same colour. The fifth dorsal spine is shown the longest, and as equalling a little more than 1½ the length of the orbit. Among Sir W. Elliot's drawings is one of this fish 1·1 inches long, it has a strong spine at the angle of the preopercle and another on the subopercle.
    Serranus præopercularis is represented by two specimens from the Persian Gulf, one 12 inches, the other 24 inches in length. The number of spines, rays and scales, as well as the form are similar to the type, but instead of black lines there are black dots along the sides rather irregularly disposed, and least numerous in the larger specimen. If, however, the smaller one be examined, the radiating lines from the eye, although indistinct, are still perceptible, giving a certain clue to what the markings had been in the young, or those of the young of the *S. morrhua*.

Page 22.   SERRANUS ANGULARIS. Add synonym.
    Perca areolata Forsk. p. 42, is identified as this species by Klunzinger.

Page 23.   For SERRANUS GRAMMICUS read S. LATIFASCIATUS. Add to synonymy.
    Serranus latifasciatus, Schlegel, Fauna Japon. Pisces, p. 6; Day, Proc. Zool. Soc. 1888, page 259.
    Having been shown the types of Schlegel's fish at Leyden, I found the two specifically identical.
    The earlier stages of growth in all fishes are interesting, more especially as they may be one means of deciding the original forms from which certain genera have probably been developed. It is, therefore, very desirable that all such should be recorded as discovered, even if merely as an incentive to further research. In 1867, 1 obtained

some small specimens of a remarkable percoid form at Madras that I termed *Priacanthichthys Madraspatensis*, but which are the young of *Serranus latifasciatus*. I have figured below in outline the head and contiguous parts of the body three times the natural size, from a specimen 1·6 inches long.

In this instance we see a preoperculum with a strong serrated spine at its angle somewhat as observed in the genus *Priacanthus*, the development of which, and the rate at which it becomes atrophied with age being most probably factors in the amount of protuberance at that spot in older fish. There is likewise a small spine on the sub-opercle, while it is very peculiar that the ventral spine is strongly serrated internally similar to a siluroid.

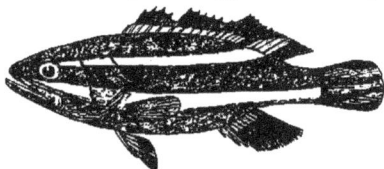

This figure is from a specimen sent by Mr. Thurston from Madras, and delineated the natural size, 3 inches long, but in which the fifth dorsal spine is abnormally shortened, giving somewhat the appearance of two spinous dorsal fins. The serrated spine at the angle of the preopercle is now only represented by coarse denticulations, the ventral spine is smooth, and the fish has assumed the form it subsequently retains. The colours are not yet similar to those of the adult, the bands along the body change as described under *Serranus morrhua* and the black bands on the caudal fin are eventually represented by more or less black spots (see Plate V, fig. 4).

SERRANUS POLLENI.

Epinephelus polleni, Bleeker, Fish. Reunion Versl. Kon. Akad. Wet. 2nd Rks. ii, p. 336, and Poisson Madagascar et Reunion, p. 19, t. vii.

B. vii, D. $_{Y6}^{9}$, P. 17, V. 1/5, A. 3/9, C. 17, L. l. 115-120.

Length of head 3½, of caudal fin 7, height of body 3¾ in the total length. *Eyes*—diameter 6 in the length of the head, 1½ diameters from the end of the snout. Preopercle serrated, most coarsely so along its lower edge. The posterior edge of the maxilla reaches to somewhat behind the hind edge of the orbit. *Teeth*—canines somewhat large. *Fins*—dorsal spines increasing in length to the fourth, the posterior ones four-fifths as high as the rayed portion of the fin, which latter is rounded posteriorly. Second anal spine somewhat the longest. Caudal fin slightly rounded. *Scales*—9 or 10 rows between the base of the sixth dorsal spine and the lateral-line. *Colours*—closely approaching those of *Serranus boenack*, being generally reddish brown, with 5 or 6 narrow blue lines on the head passing backwards, 10 or 12 horizontal blue lines along the body, those in the upper third being interrupted and contorted, some ending on the soft portion of the dorsal fin, while others similarly end on the caudal and anal fins. A horizontal narrow blue band along the centre of the dorsal fin, and a narrow blue edging to the soft portion, 2 or 3 blue bands taking a semi-horizontal direction on the anal fin. Caudal externally edged with reddish brown.

5 π 2

An example from the Isle of France exists in the National Museum; another is in the Paris Museum, received from M. Lienard, of the Mauritius, and the coloured figure of a Madras specimen termed *Sembarra punnee*, Pam., exists in the late Sir Walter Elliot's collection made at Madras.

Page 24. SERRANUS GUTTATUS.

It has been my belief that this is a variety of the *S. hemistictus*, the latter wanting the cross bands. *S. guttatus* was figured in colours in the *Fische de Sulsee* by the late Mr. Ford, and as the specimen was superior to mine I requested him to employ the same fish for my uncoloured representation. He did so, but the bands had faded, and now the Südsee figure is referred to as *S. guttatus*, mine as *S. hemistictus*. As both were from the identical specimen by the same artist, it affords an additional argument that they are varieties of one species.

Page 25. SERRANUS LEOPARDUS. Add to synonymy.

Serranus sexmaculatus, Rüpp. Atl. Fische, p. 107.
Serranus zanana, Cuv. and Val. ii, p. 339.

Page 27. VARIOLA LOUTI. Add to synonymy.

Serranus flavimarginatus, Rüpp. Atl. p. 109.
Serranus novemcinctus, Kner, Novara Fische, p. 17, t. ii, f. 1.

Page 27. For ANTHIAS MULTIDENS read APRION PRISTOPOMA. Add synonyms.

Pristipomoides typus, Bleeker, Sumatra, p. 575.
Dentex pristopoma, Bleeker, Celebes, p. 216.
Mesoprion dentex, Bleeker, Enum. Pisc. p. 20.
Lutjanus dentex, Bleeker, Amboina, p. 278.
Chætopterus pristipoma, Bleeker, Chœtop. p. 83, and Fish. Madagascar, t. x.
Anthias multidens, Day, Fish. India, p. 27, pl. vii, f. 4.
Aprion pristipoma, Bleeker, Atl. Ich. viii, p. 79; Perc. t. lviii, f. 3.
Centopristis pristopoma, Klunzinger, Fisc. Roth. Meer, p. 16.

Add Genus. Apharous, Cuv. and Val.

Branchiostegals seven, pseudobranchiæ. Body oblong. Cleft of mouth wide and oblique, the lower jaw the longer. Preopercle and preorbital entire. Canines absent, palate edentulous. A single dorsal fin, with the front portion the highest, spines (10-12) feeble: caudal forked. Scales small. Air-bladder simple. A respiratory cavity behind the branchiæ. Pyloric appendages few.

Page 27. APHARFUS RUTILANS.

Aphareus rutilans, Cuv. and Val. vi, p. 490; Rüppell, N.W.F. p. 121; Bleeker, Amboina, p. 52, and Atl. Ich. vii, Perc. t. xxi, f. 2; Günther, Catal. i, p. 386; Klunz. F.R.M. p. 45.
Aphareus furcatus, Bleeker, Atl. Ich. viii, p. 80.

B. viii, D. $\frac{10}{3-14}$ P. 16, V. 1/5, A. $\frac{3}{8}$, C. 17, L. l. 65-70, Cœc. pyl. 5.

Length of head 4, of caudal fin 3½ to 3⅔, height of body 4½ to 5 in the total length. *Eyes*—diameter 4 in the length of the head, 1¼ diameters from the end of the snout and 1 apart. *Teeth*—anteriorly in two or three rows, in a single row posteriorly. *Fins*—fourth and fifth dorsal spines the highest in the fin; last dorsal and anal rays prolonged to twice the length of the penultimate ones; caudal deeply forked. *Colours*—rosey, darkest along the back, with a yellow blotch between each ray of the dorsal fin near their bases; caudal lobes with dark tips.

Bleeker considered this species to be identical with *Labrus furcatus*, Lacép., or *Aphareus cœrulescens*, Cuv. and Val.

Habitat.—Red Sea to the Malay Archipelago. Obtained at Ceylon by Mr. Haly.

Page 28. Add GRAMMISTES PUNCTATUS.

Cuv. and Val. vi, p. 504; Günther, Fische d. Südsee, 1875, p. 11, t. ii, f. B; Bleeker, Fish. Madagascar, 1874, p. 24, t. xiii, and Atl. Ich. vii, p. 69, Perc. t. lix, f. 5.

B. vii, D. 7/$\frac{1}{14}$, P. 16, V. 1/5, A. 11, C. 17.

Length of head 3¾, of caudal fin 7, height of body 4 in the total length. *Eyes*—high up, diameter 5½ in the length of the head, about ⅓ a diameter apart. Lower jaw the longer. Vertical limb of preopercle with spinate denticulations: three spines on opercle. A barbel, rather longer than one diameter of the eye, at the symphysis of the lower jaw.

Teeth—generic. *Scales*—small, imbedded in mucus. *Colours*—grayish-brown, with small white dots.

A third species, with 7 or 8 dorsal spines and a more elongated body, has been found at the Seychelles, and appears to be *O. compressus*, Lienard.

Page 33. For LUTJANUS BENGALENSIS read *L. kasmira.*

Forskal's species appears to be identical with *Holocentrus Bengalensis*, Bloch, but the variety he mentions with a black lateral blotch is *H. quinquelinearis*, Bloch, and is certainly a distinct species, the latter differing, irrespective of colouring, in many important points from the former, for its preopercular notch is deeper, its eye larger, its snout more rounded, &c. In some specimens the bands on the side are red, not blue.

Page 37. LUTJANUS ARGENTIMACULATUS. Add synonym.

Mesoprion garretti, Günther, Fische Südsee, p. 15, t. xiii, f. B.

Page 40. For LUTJANUS QUINQUELINEATUS read CÆRULEOLINEATA. Add synonym.

Mesoprion cœruleolineata, Klunz. F.R.M. p. 15. Erase synonym

Holocentrus quinquelineatus, Bl. Schn, as this appears to have been a misprint for *H. quinquelinearis*, and referred to Bloch's figure No. 239. The species here described is identical with one of Bloch's specimens thus marked in the Berlin Museum.

Page 41. LUTJANUS FULVIFLAMMA. Omit from synonyms.

Sparus antika doondiawah, Russell, Fish. Vizag. i, p. 76, pl. 98.
Mesoprion unimaculatus, Quoy and Gaim. Voy. Freyc. p. 304, &c.
     ,,   *anrolineatus*, Cuv. and Val. iii, p. 496.
     ,,   *Russellii*, Bleeker, Perc. p. 41.
Lutjanus notatus, Bleeker, Ternat. p. 238.
Genyoroge notata, Cantor, Catal. p. 12.
Mesoprion ehrenbergii (Peters) Boulenger, Pro. Zool. Soc. 1887, p. 665.

Page 42. LUTJANUS FULVIFLAMMA, var. RUSSELLII. Add synonyms as omitted from last species.

Lutjanus russellii, Bleeker, Atl. Ich. viii, p. 71, Perc. t. xxii, f. 2.

Page 42. Add LUTJANUS NIGRA.

Sciæna nigra, Forsk. Desc. Anim. p. 47 ; Gmel. Linn. p. 1300.
Lutjanus nigra, Bl. Schn. p. 326.
Diacope nigra, Cuv. and Val. ii, p. 431; Rüpp. N.W. Fische, p. 93, t. xxiv, f. 1 ; Klunz. Fische, R. M. p. 11.
Proamblys niger, Gill, Proc. Ac. N. Sc. Phil. 1862, p. 236.
(*Young.*)
Diacope macolor, Cuv. and Val. ii, p. 415 ; Less. Mém. Soc. Hist. Nat. iv, p. 409, and Voy. Coq. Zool. ii, p. 230, pl. xxii, f. 2.
Mesoprion macolor, Bleeker, Celebes, iii, p. 753.
Genyoroge macolor et nigra, Günther, Catal. i, p. 176, Fish. Zanzibar, p. 14.
Macolor typus, Bleeker, Amboina, Ned. T. Dierk. ii, p. 277.
Lutjanus macolor, Bleeker, Atl. Ich. viii, p. 75, Perc. t. lxv, f. 3.

B vii, D. $\frac{10}{13-14}$, P. 17, V. 1/5, A. $\frac{3}{13-14}$, C. 17, L. l. 45-50, Cœc. pyl. 4.

Length of head 3⅓, of caudal fin 4¼ to 4¼, height of body 3 to 3¼ in the total length. *Eyes*—3¼ to 4 diameters in the length of the head, 1 diameter from the end of the snout, and 1 apart. Upper profile of head very convex. Lower jaw the longer. Vertical limb of preopercle with a deep notch to receive a large interopercular knob, and its lower edge serrated. *Fins*—dorsal and anal with their soft parts pointed, pectoral long, reaching the anal. Caudal emarginate. *Colours*—adult, of a grayish-black, *immature*, dark purplish, nearly black (belly bluish), with several light spots along the base of the dorsal fin. A light band along the middle of the body and tail fin. Another from the eye over the jaws, and a third down the opercle. Fins dark, the posterior ends of dorsal and anal fins light-coloured. Caudal lobes tipped with white.

Bleeker considered that *Lutianus nigra* to be distinct from *L. macolor.*

Habitat.—Red Sea, East Coast of Africa, Navigator Islands to the Malay Archipelago, and Mr. Haly in 1887 had an example sent from the Maldives to the Ceylon Museum.

Page 48. 1. PRIACANTHUS BLOCHII. Add synonyms.

Sciæna hamruhr, Forsk. Des. An. p. 45.
Anthias hamruhr, Bl. Schn. p. 307.
Priacanthus hamruhr, Cuv. and Val. iii, p. 104; Günther, Catal. i, p. 219 ; Bleeker, Atl. Ich. vii, p. 13, Perc. t. lxxv, f. 3.

Priacanthus macracanthus, Cuv. and Val. iii, p. 108 ; Günther, Catal. i, p. 220.
　　　,,　　*fax*, Cuv. and Val. vii, p. 473; Günther, Catal. i, p. 220.

Page 48.　2. PRIACANTHUS HOLOCENTRUM, page 746.　Add synonym.
Priacanthus tayenus, Richards. Ich. China, p. 237; Günther, Catal. i, p. 221 ; Bleeker, Atl.
　　Ich. vii, p. 12, Perc. t. lxxi, f. 4.
Priacanthus schmittii, Bleeker, Sumatra, p. 572 ; Günther, Catal. i, p. 220.

Page 51.　AMBASSIS RANGA.　Add synonym.
Ambassis notatus, Blyth, P. Asi. Soc. Beng. 1860, p. 138 (not *A. baculis*).

Page 55.　Add AMBASSIS MYOPS.
Ambassis myops, Günther, P. Z. S., 1871, p. 655.

　　　　B. vi, D. 7/$\frac{1}{3}$, P. 12, V. 1/5 A. $\frac{3}{8}$, C. 17, L. l. 29, L. tr. 4/9.

　　　Length of head 4, of caudal fin 4$\frac{1}{4}$, height of body 3$\frac{1}{4}$ in the total length. *Eyes*—
diameter one-third of the length of the head, 2/3 of a diameter from the end of the snout,
and the same distance apart.　Lower jaw the longer.　Cleft of mouth very oblique: the
maxilla reaches to beneath the front edge of the orbit.　Preorbital with seven strong
teeth along its lower edge : a spine at the posterior-superior angle of the orbit.　Vertical
limb of preopercle entire : its horizontal double edge serrated the lower most coarsely so.
Sub- and inter-opercles entire.　*Teeth*—villiform in jaws, vomer, and palate, a small central
band at the root of the tongue.　*Fins*—second spine of the dorsal longest and equal to 4$\frac{3}{4}$
in the total length, and 2/3 the height of the body below it.　Ventrals reach the vent :
pectoral reaches to above the third anal spine, which latter is longer and weaker than the
second, but half shorter than the third dorsal spine.　*Scales*—two to three rows along the
cheeks.　*Lateral-line*—curves to near the middle of the soft dorsal, when it becomes straight,
it is uninterrupted.　*Colours*—silvery, with a burnished lateral band.　Interspinous
membrane between the second and third dorsal spines spotted with black.

　　　Habitat.—Sea at Madras, from which Mr. Thurston has sent me one specimen 4 inches
long, to the Malay Archipelago and Cook's Islands.

Page 59.　APOGON ENDEKATÆNIA.　Omit species and unite with A. FASCIATUS, p. 60.
Page 61.　Add APOGON THURSTONI.

　　　　B. vii, D. 7/$\frac{1}{3}$, P. 14, V. 1/5, A. $\frac{3}{8}$, C. 17, L. l. 26, L. tr. 2/6$\frac{1}{2}$.

　　　Length of head 3$\frac{1}{4}$, of caudal fin 5$\frac{1}{4}$, height of body 2$\frac{3}{4}$ in the total length. *Eyes*—
diameter 3$\frac{1}{4}$ in the length of the head, $\frac{3}{4}$ of a diameter from the end of the snout, and 1
apart.　A very slight rise from the snout to the base of the dorsal fin.　Snout a little
elevated : upper jaw slightly the longer, and extending posteriorly to below the last third
of the orbit.　Both limbs of the preopercle serrated, the vertical one finely and evenly, the
angle rather coarsely and the lower limb more irregularly : shoulder scale serrated.　*Teeth*
—villiform ones in jaws, also present on vomer and palate.　*Fins*—Dorsal spines strong,
the two first short, the third slightly the longest, and equal in length to the head behind
the middle of the eyes, and nearly half the height of the body below it : the rays of the
second dorsal as long as the longest dorsal spine and one-fourth longer than those of the
anal fin.　Pectoral reaches to above the anal spines, and the ventral nearly as far.　Caudal
somewhat square at its extremity.　*Lateral-line* very slightly curved, becoming straight
on the free portion of the tail: its tubes simple with a basal expansion on each side.
Colours—greyish, darkest along the back and a dark band behind the base of the second
dorsal fin : an oval black spot nearly as large as the orbit and surrounded by a narrow
yellow ring exists below the lateral line and under the first dorsal fin.　Vertical fins black,
caudal yellowish.

　　　Habitat.—Madras, from which Museum I have been lent by Mr. Thurston, a specimen
3 inches long.

Page 62.　APOGON BIFASCIATUS.　Add synonym.
Apogon maximus, Boulenger, P. Z. S. 1887, p. 655.

　　　Some very fine examples, in which the black spots are unusually large, were received
from the Persian Gulf and thus named.　The third and fourth dorsal spines are only half
the length of the head, the eye is naturally smaller than in recorded specimens, and the
maxilla extends to below the middle of the eye.　These fish reach to about 10 inches
in length.

Page 63.　APOGON ELLIOTI.　Add synonym.
Apogon arafura, Günther, Challenger Shore Fishes, 1880, p. 38, pl. xvi, f. c.

Page 64. APOGON MACROPTERUS. Add synonym.

Apogon lineolatus (Ehr.) Cuv. and Val. ii, p. 160; Rüpp. Atl. p. 47, t. xii, f. 2.

Page 65. Add APOGON TICKELLI.

Apogon pœcilopterus, Cantor, Catal. p. 2 (not Cuv. and Val.).

B. vii, D. 6/$\frac{1}{13}$, P. 13, V. d/5, A. $\frac{2}{8}$, C. 15, L. l. 24 (26) L. tr. 3/8.

Length of head 3 to 3½, of caudal fin 5½, height of body 3½ in the total length. *Eyes*—diameter ¼ of length of head, nearly 1 diameter from the end of the snout, and ¾ to 1 diameter apart. Lower jaw very slightly the longer. The maxilla reaches to slightly behind the hind edge of the eye. The posterior edge of the preopercle finely serrated except in a small portion of its lower part. A considerable rise from the snout to the base of the first dorsal fin. *Fins*—first dorsal spine one-third the length of the second, which is equal to the third and about 2½ in the length of the head; second dorsal somewhat higher than the first. Caudal rounded. *Scales*—finely ctenoid. *Colours*—pale horn above and below, with a slight golden tinge on the opercles: caudal and ventral both having a dark hind edge. A round black spot at the root of the caudal fin.

Habitat.—Col. Tickell procured two examples at Akyab (see figure 4·2 inches long, "scale 10/16," p. 215, MSS.) and it seems to be identical with Cantor's fish.

Page 66. For CHEILODIPTERUS LINEATUS read C. MACRODON.

Omit synonyms *Perca lineata*, Forsk., *P. arabica*, Linn., *Cheilodipterus lineatus*, Lacép., and *C. arabicus*, Cuv. and Val.

Add *Paramia macrodon*, Bleeker, Atl. Ich. vii, p. 105.

Page 66. Add 3. CHEILODIPTERUS LINEATUS, also synonym omitted from last species.

Page 71. Genus *Datnia* to be included with genus *Therapon*.

Page 72. Plate xviii, fig. 8, for *P. nageb* read *P. stridens*.

Page 80. Add DIAGRAMMA CUVIERI.

Bodian cuvieri, Bennett, Fish. Ceylon, p. 13, pl. xiii.
Diagramma sebæ, Bleeker, Sciænidæ, p. 24.
Plectorhynchus sebæ, Bleeker, Atl. Ich. Perc. t. xxvii, f. 3.
Diagramma lessonii, Günther, Catal. i, p. 329, and Fische Südsee, p. 28, t. xxiii (not Cuv. and Val.).
Diagramma cuvieri, Playfair, Fish. Zanzibar, p. 28.
Plectorhynchus cuvieri, Bleeker, Atl. Ich. viii, p. 21.

B. vii, D $\frac{12}{22\cdot13}$, P. 18, V. 1/5, A. $\frac{3}{8}$, C. 15, L. l. 70, L. tr. 11/30.

Length of head 3½ to 4, of caudal fin 7, height of body 3½ in the total length. *Eyes*—diameter 2½ to 3½ in the length of the head, 1½ diameters from the end of the snout, and 1 apart. The maxilla reaches nearly to beneath the front edge of the eye. Vertical limb of preopercle serrated. *Fins*—dorsal spines slightly higher than the rays, the second to the fourth being of about the same length, and the longest in the fin, while each is about equal to one-third the height of the body; second anal spine the longest and strongest. *Scales*—ctenoid. *Colours*—silvery with horizontal greyish or brownish bands, the upper of which are wider than the ground colour, these bands unite anteriorly over the nape and snout, while the upper ones end posteriorly at the base of the dorsal fin. Fins yellowish, the dorsal, caudal and anal with some dark bands and spots and dark outer edges.

A specimen 7½ inches long received from Madras through the kindness of Mr. Thurston, has D $\frac{12}{14}$, which is very interesting, as showing how great a variation in the number of spines and rays may exist, for the usual numbers are D $\frac{12}{22\cdot13}$.

Habitat.—From the East Coast of Africa, to Ceylon, Madras, and the Malay Archipelago to 380^m: and in the British Museum to 14¾ inches.

Page 81. DIAGRAMMA GRISEUM. Add synonym.

Diagramma jayakari, Boulenger, P. Z. S. 1887, p. 656.

This differs from the types in having one more spine and ray in the dorsal fin, or D 13/22, but Mr. Thurston has lately sent me a specimen from Madras with D 12/22. Some error occurred in Mr. Boulenger's description, for if "the greatest depth of the soft dorsal equals the length of the longest spine, or seven-eighths the depth of the body," this fin would be enormously developed. However, we are also informed that the longest dorsal spine is " not quite one-third the length of the head," and it is manifestly improbable that any Diagramma would have the length of its head equalling nearly three times the height of the body when that height is " thrice and two-fifths in the total length." In fact the form is similar to that figured as *D. griseum*, C.V.

In the "Fishes of Zanzibar" it was pointed out that *D. griseum* was subject to variations in colour, and one was figured showing four whitish curved cross bands. There is no genus of Asiatic marine fishes with more variation in the colour of individual specimens and local races than shown in that of *Diagramma*. While I stated that in the young some sinuous and narrow light blue lines exist over the snout and cheeks, and also several sinuous blue lines taking an oblique direction from the head upwards, and which extend to nearly the length of the body.

Page 92.   SYNAGRIS JAPONICUS.   Add synonym.

> *Dentex filamentosus*, Steind. Sitz-Bert. Akad. Wien. 1868, p. 976.
> „   *blochii*, Bleeker, Atl. Ich. viii, p. 90, Perc. t. lii, f. 4.

Page 93.   For SYNAGRIS NOTATUS read S. TÆNIOPTERUS.   Add synonym.

> *Dentex tæniopterus*, Cuv. and Val. vi, p. 246 ; Bleeker, Atl. Ich. viii, p. 83, Perc. t. lvi, f. 5.

Page 96.   For DATNIOIDES POLOTA read D. QUADRIFASCIATUS.   Add synonyms.

> *Chætodon quadrifasciatus*, Sevastion, Mem. Acad. St. Peters. 1809, i, p. 448, t. xviii.
> *Datnioides quadrifasciatus*, Bleeker, Atl. Ich. viii, p. 32, Perc. t. xxvii, f. 1.

Page 97.   GERRES SETIFER.   Add synonym.

> *Gerres altispinis*, Günther, Introd. Study of Fish, p. 388, and fig. 150.

Page 106.   For CHÆTODON GUTTATISSIMUS read C. MILIARIS.   Add synon.

> *Chætodon guttatissimus*, Günther, Fische Südsee, i, p. 46, t. xxxv, f. A.
> „   *citrinellus*, Cuv. and Val. vii, p. 27 ; Günther, l. c. p. 47, t. xxxv, f. B.
> *Tetragonoptrus miliaris*, Bleeker, Atl. Ich. ix, p. 39, t. 377, Chæt. t. xv, f. 3.

Page 107.   For CHÆTODON VITTATUS read C. TRIFASCIATUS.   Add synonyms.

> *Chætodon tau nigrum*, Cuv. and Val. vii, p. 38 (*young*).
> *Citharoedus vittatus*, Kaup, Arch. Nat. 1860, p. 142.
> *Tetragonoptrus trifasciatus*, Bleeker, Atl. Ich. ix, p. 35, t. 377, Chætod. t. xv, f. 1.

Page 108.   For CHÆTODON LUNULA read C. FASCIATUS.   Add synonyms.

> *Chætodon fasciatus*, Forsk. Descrip. Anim. p. 59.
> „   *flavus*, Bl. Schn. p. 225.
> „   *ocellatus*, Bleeker, Timor, p. 212.
> „   *wiebeli*, Kaup, Chætod. i, p. 126.
> *Tetragonoptrus fasciatus*, Bleeker, Atl. Ich. ix, p. 41, t. 374, Chætod. t. xii, f. 2.

Page 109.   For CHÆTODON OLIGACANTHUS read C. OCELLATUS.   Add synonym.

> *Parachætodon ocellatus*, Bleeker, Atl. Ich. ix, p. 24, pl. 377, Chætod. t. xv, f. 4.

Page 110.   ZANCLUS CORNUTUS.   Add synonyms.

> *Chætodon canescens*, Linn. Syst. Nat. i, p. 466 (*young*).
> *Zanclus centrognathus*, Cuv. and Val. vii, p. 528 („).
> *Chætodon nudus*, Gronov. ed. Gray, p. 76.
> *Zanclus canescens*, Günther. Catal. ii, p. 493 (*young*).
> *Gnathocentrum centrognathum*, Guichen. Ann. Soc. Linn. Maine et Loire, ix, Ich. p. 4 (*young*).
> *Zanclus cornutus*, Bleeker, Atl. Ich. ix, p. 77, Chætod. t. iv, f. 1, 2.

Page 126.   Add

FAMILY—MALACANTHIDÆ, *Günther*.

Branchiostegals from five to six: pseudobranchiæ present. Gill-openings wide, the membranes united beneath the throat. Gills four, with a slit behind the fourth. Body elongated and compressed ips thick. A posterior canine tooth in the premaxillaries. Dorsal and anal fins with many rays, the first few of the former not being articulated. Ventrals thoracic with one spine and five rays. Scales small, and finely ctenoid. Air-bladder simple. Pyloric appendages absent.

Genus 1. MALACANTHUS, *Cuv.*

Cleft of mouth horizontal, with the jaws equal anteriorly. Opercle with a spine: pre-opercle entire. Eyes lateral. Villiform teeth in the jaws, having an outer band of stronger ones: palate toothless. A long continuous dorsal fin with the first four to six rays not articulated.

Habitat.—Tropical seas.

1. MALACANTHUS LATOVITTATUS.

Labrus latovittatus, Lacép. iii, p. 526, pl. xxviii, f. 2.
Tænianotus latovittatus, Lacép. iv, p. 304.
Malacanthus latovittatus, Quoy and Gaim. Voy. Astrol. iii, p. 701, pl. xx, f. 3; Günther, Catal. iii, p. 361.
Malacanthus tæniatus, Cuv. and Val. xiii, p. 327, pl. 381; Bleeker, Nat. Tyds. Ned. Ind. ii, p. 218.

D. iv-v, D. $\frac{1}{5}\frac{1}{17}$, P. 17, V. 1/5, A. $\frac{1}{17}\frac{1}{50}$, C. 17, L. l. 125.

Length of head 4, of caudal fin 9, height of body 6 to 7 in the total length. *Eyes*—high up, and situated nearly midway between the end of the snout and the posterior extremity of the opercle, diameter 7 in the length of the head: cleft of mouth does not reach to below the front edge of the orbit. *Fins*—the dorsal commences above the axil of the pectoral but does not extend on to the caudal. *Colours*—brownish with a broad black band along the side from the pectoral to the caudal fin.

Habitat.—New Guinea, Mauritius. Ceylon (Haly).

Page 134. For LETHRINUS ROSTRATUS read L. MINIATUS. Add synonyms.

Sparus miniatus (Forster) Bloch, Schn. p. 281.
Lethrinus miniatus, Cuv. and Val. vi, p. 316; Bleeker, Atl. Ich. viii, p. 121, Perc. t. xxxi, f. 3.
Lethrinus olivaceus et *waigiensis*, Cuv. and Val. vi, pp. 295, 297.
  „   *acutus*, Klunz. Fis. R. Meeres, p. 38, t. vii, f. 1.

Page 138. For SPHÆRODON HETERODON read S. GRANDOCULIS. Add synonyms.

Sciæna grandoculis, Forsk. p. 53.
Chrysophrys grandoculis, Cuv. and Val. vi, p. 134.
Lethrinus latidens, Cuv. and Val. vi, p. 316.
Sphærodon grandoculis, Rüppell, N. W. Fische, p. 113, t. xxviii, f. 2.
  „   *latidens*, Kner, Novara Fische, p. 83, t. iv, f. 1.
Monotaxis grandoculis, Bleeker, Atl. Ich. viii, p. 105, Perc. t. xxi, f. 1.

Page 138. PAGRUS SPINIFER. Add synonym.

Pagrus ruber, Boulenger, Proc. Zool. Soc. 1887, page 658.

Mr. Boulenger has instituted a new species from the Persian Gulf having "a protuberance between and in front of the eyes;" of the dorsal spines the "third longest, compressed and curved, its length one-third to one-fourth the depth of the body," being apparently considered sufficient to characterize it. In Cuv. and Val. we are told that in *Pagrus spinifer* the third dorsal spine is 2/3 the height of the body, and the fourth about the same length. If, however, a large number of specimens are brought together it becomes at once apparent that this difference in the length of the dorsal spines is almost entirely owing to the age and size of the example. In two young specimens from Sind, each 3 inches long, the filamentous prolongation reached the caudal fin, but as age increases the comparative length of these filaments diminishes. The largest of Mr. Boulenger's two specimens is 19 inches long, its third dorsal spine is 1·8 inches or 10⅝ in the total length; while the smaller example is 13 inches long, and its third dorsal spine 1·7 inches long or 7⅟⅟ in the total. In two small specimens, 7 and 7·4 inches respectively in length sent to the British Museum by Colonel Playfair, the length of the prolonged rays is absolutely greater than in the larger examples. In a Madras specimen 4·8 inches long the third dorsal spine is 2 inches long, or 2¼ in the total length; and in a larger example 9 inches long the third dorsal spine is 1·8 inches long, or 1/5 of the total length, while the frontal protuberance is well developed. I figured an intermediate sized one in which this spine was about 3¼ in the total length or 2/3 of the height of the body. It is no doubt true that in the young considerable variations are seen in the length of these rays, which prolongations become absorbed with age. The two types of *P. ruber* are somewhat large specimens, but if we examine the foregoing figures we see as follows respecting the third dorsal spine, at 19 inches it equals 10⅝, at 13 inches 7⅟⅟, at 9 inches 5, at 7 inches 3¼, at 4·8 inches 2¼ in the total length.

Page 140.   CHRYSOPHRYS DATNIA.   Add synonyms.

Sparus hasta, Bleeker, Revis. 1876, p. 9, t. iii, and Atl. Ich. viii, p. 108, Perc. t. lxvii, f. 3.

Page 140.   CHRYSOPHRYS BERDA.   Add synonyms.

Sparus datnia, Bleeker, Revis. 1876, p. 5, t. ii, and Atl. Ich. viii, p. 109, Perc. t. lxxvii, 4 f.

Page 141.   CHRYSOPHRYS CUVIERI.   Add synonyms.

Sparus cuvieri, Bleeker, Over. Vers. en Meded. der Konig. Akad. v. Weter. 1877, c. fig.

Page 142.   For CHRYSOPHRYS HAFFARA read C. ARIES.   Omit synonymy and insert,

Sparus haffara, Forsk. &c.

Page 143.   For PIMELEPTERUS FUSCUS read P. WAIGIENSIS.   Omit synonyms and insert.

Xyster fuscus (Comm.) Lacép. V. pp. 484, 485.
Pimelepterus fuscus, Cuv. and Val. vii, p. 264, &c.

Page 145.   For CIRRHITES FASCIATUS read CIRRHITICHTHYS FASCIATUS.

Having obtained some small specimens of this fish from Madras, I find teeth present on the vomer and palatine bones.

Page 150.   SCORPÆNOPSIS GUAMENSIS.   Add synonyms.

Scorpæna rubropunctata (Ehren.) Cuv. and Val. iv, p. 324.
Sebastes minutus, Cuv. and Val. iv, p. 348.
Scorpæna chilioprista, Rüpp. N. W. F. p. 107, t. xxvii, f. 2.
    „   *guamensis*, Günther, Fische Südsee, p. 74, t. 56, f. B.
Sebastopsis polylepis, Gill, Proc. Ac. Nat. Sc. Phil. 1862, p. 278; Bleeker, Scorp. 1873, p. 21, t. iv, f. 2, and Atl. Ich. Scorp. t. v, f. 1.

Page 150.   For SCORPÆNOPSIS OXYCEPHALA read S. LEONINA.   Add synonym.

Scorpæna leonina, Richardson, Ich. China, p. 216.

Page 154.   For PTEROIS CINCTA read P. RADIATA.

Page 163.   For Genus PSEUDOSYNANCEIA read LEPTOSYNANCEIA, Bleeker.

This latter genus is said *to possess vomerine teeth.*

Page 169.   MYRIPRISTIS BOTCHE.   Add synonym.

Myripristis murdjan var. *adusta*, Günther, Fische Südsee, p. 92, pl. lxii.

Page 173.   HOLOCENTRUM SAMMARA.   Add synonym.

Holocentrum platyrhinum, Klunz. Synopsis Fische R. M. p. 725.

Page 175.   For PEMPHERIS MANGULA read P. MALABARICA.

Omit synonyms and add
Pempheris malabarica, Cuv. and Val. vii, p. 308.

Page 175.   For PEMPHERIS MOLUCCA read P. RUSSELLII.

Omit synonyms.   Add
Sparus mangula-kutti, Russell, Fish. Vizag. ii, p. 10, pl. xiv.
Pempheris mangula, Bleeker, Atl. Ich ix, Pemph. t. i, f. 2 (not Cuv. and Val.).
?   „   *rhomboideus*, Kossm. and Raüber, Fis. R. M. p. 18, t. i, f. 4.
    This species is identical with Russell's fish but not with *P. mangula* C.V., a form figured in Günther's Fische d. Südsee, t. lix, f. B; whereas Klunzinger's *P. mangula* differs again from both species.

Page 182.   UMBRINA SINUATA.   Add synonym.

Umbrina striata, Boulenger, P. Z. S. 1887, p. 660.

Page 200.   TRICHIURUS MUTICUS.   Add synonym.

Trichiurus cristatus, Klunz. F. R. M. p. 121, t. xiii, f. 5 (head).

Page 201.   TRICHIURUS SAVALA.   Add synonym.

? *Trichiurus auriga*, Klunz. F. R. M. p. 121, t. xii, f. 1.

Page 204.   ACANTHURUS TENNENTII.

    Col. Tickell, MSS., p. 297, has a figure of a form from Arrakan very similar to this species, but with 8/25, A. 3/23, which he termed *A. tristis*. *Colours*—slate-gray and slightly ciuereous along the back. Dorsal fin olivaceous along its base: caudal blackish along its centre, nearly white externally. Ventrals whitish, externally black, an irregular black band extends from the upper edge of the orbit across the top of the operclo to the base of the pectoral fin which it crosses.

Page 205. For ACANTHURUS MATA read A. GAHM. Add synonyms.

Acanthurus gahm, Forsk. p. 64; Cuv. and Val. x, p. 219; Günther, Fische Südsee, i, p. 113, t. lxxiv.
Acanthurus nigricans, Rüppell, Atl. p. 27.
    „   matoides, Günther, Catal. iii, p. 330.
    „   annularis, Cuv. and Val. x, p. 209.
    „   Blochii, Cuv. and Val. x, p. 209; Günther, Fische Südsee, i, p. 109, t. lxix, f. B.
    „   melanurus, Cuv. and Val. x, p. 240; Günther, Catal. iii, p. 346 (young).
    „   argenteus, Quoy and Gaim. Voy. Uranie, p. 372, t. lxiii, f. 2; Günther, Catal. iii, p. 346 (young).
    „   xanthopterus, Cantor, Catal. Malayan. Fish, p. 209, pl. iv.

Page 207. ACANTHURUS STRIOSUS. Add synonyms.

Acronurus lineolatus, Klunz. F. R. M. Synopsis, ii, p. 511.
Acanthurus striatus, Günther, Fische Südsee, i, p. 116, t. lxxix, f. B. (? ? Quoy and Gaim. Voy. Uranie, p. 373, pl. lxiii, f. 3).

Page 214. CARANX KURRA. Add synonym.

Decapterus Russellii, Klunz. F. R. M. page 91.

Page 214. CARANX MELAMPYOUS. Add synonyms.

Caranx stellatus, Eyd. and Soul. Voy. Bonite, Poiss. p. 167, t. iii, f. 2.
    „   bixanthopterus, Rüpp. N. W. F. p. 49, t. xiv, f. 2.

Page 216. CARANX HIPPOS. Add synonyms.

Caranx flavo-cœrulens, Schlegel, Fauna Japon. Pisces, p. 110, t. lix, f. 2.
    „   parapistes, Richards. Voy. Erebus and Terror, p. 136, pl. lviii, f. 6, 7.
Carangus marginatus, Gill, Proc. Phil. Acad. 1863, p. 166.
Caranx caninus, Günther, Trans. Zool. Soc. vi, p. 432.

Page 217. CARANX FERDAU. Add synonyms.

Carangoides hemigymnostethus, Bleeker, Mackrel, p. 61.
Caranx venator, Playfair, Fish. Seychelles, P. Z. S. 1867, p. 859, fig. 2.

Page 221. CARANX MALABARICUS. Add synonyms.

Carangoides telamparoides, Bleeker, Makrel, pp. 69, 91.
Caranx impudicus, Klunz. F. R. M., p. 99.

Page 223. CARANX NIGRESCENS. Add synonym.

Caranx jayakari, Boulenger, P. Z. S. 1887, p. 661.

Page 226. CARANX SPECIOSUS. Add synonyms.

Caranx rüppellii, Günther, Catal. ii, p. 445.
    „   edentulus, All. and Macleay, 1877, p. 327.

Page 228. SERIOLICHTHYS BIPINNULATUS. Add synonyms.

Seriola pinnulata, Poey, Mem. ii, 1858.
Elagatis pinnulatus, Gilb. in F.-W. Fish. N. America, 1883, p. 446.

Page 230. CHORINEMUS SANCTI-PETRI. Add synonyms.

? Lichia toloopurah, Rüpp. Atl. p. 91.
Chorinemus tol, Kner, Novara Fish. p. 162.
    „   toloo, Klunz. F. R. M. Synopsis, p. 447 (not Cuv. and Val.).
    „   moadetta, Klunz. F. R. M. p. 105 (not Cuv. and Val.).

Page 230. CHORINEMUS MOADETTA. Add synonym.

Chorinemus mauritiana, Cuv. and Val. viii, p. 382.

Page 231. CHORINEMUS LYSAN. Add synonym.

Chorinemus orientalis, Schlegel, Fauna Japon. Pisces, p. 106, t. lvii, f. 1.

Page 233. TRACHYNOTUS RUSSELLII. Add synonym.
Trachynotus coppingeri, Günther, Fish. Alert Expedition, 1881-2, p. 29, pl. iii, f. A.

Page 234. TRACHYNOTUS OVATUS. Add synonym.
Trachynotus kennedyi, Steind. SB. Ak. Wein. lxxii, p. 75, f. 9.

Page 237. PSENES JAVANICUS. Add synonym.
Psenes guamensis, Günther, Fische Südsee, ii, p. 145, t. xci, f. 100.

Page 244. Add GAZZA ARGENTARIA.
Zeus argentarius (Forster) Bloch, Schn. p. 96; Forster, Descr. Anim. p. 288.
Gazza tapeinosoma, Bleeker, Sumatra, p. 260.
      „   argentaria, Günther, Catal. ii, p. 506, Fische Südsee, ii, p. 144, pl. xci, f. B.; Klunz.
    F. R. M. p. 108.
Equula dentex, Peters, Fish. Moss. p. 246 (not C. V.).

   Length of head 3¼ to 4, of caudal fin 5, height of body 2¼ to 2½ in. of the total length.
Eye—diameter ⅓ of the length of the head, 2/3 of a diameter from the end of the snout.
Teeth—canines of moderate size. Fins—first dorsal higher than the second. Colours—
body grayish, with some dark lines passing along the rows of scales, light-coloured on the
chest; dorsal, anal, and ventral fins nearly black, caudal of a dull yellow.
   Habitat.—Red Sea, Madras to the Malay Archipelago. A coloured figure named Psani
paré, Tamil, exists among the late Sir W. Elliot's drawings.

Page 250. SCOMBER MICROLEPIDOTUS. Add synonyms.
Scomber loo, Cuv. and Val. viii, p. 52.
     „   moluccensis, Bleeker, Amboina, p. 40.

Page 251. Add SCOMBER JANESABA.
Scomber pneumatophorus minor, Schleg. Fauna Japon. Pisces, p. 94, pl. xlvii, f. 2.
     „   janesaba, Bleeker, Japan. p. 406, and Verh. Bat. Gen. xxvi Japan. p. 96;
    Günther, Catal. ii, p. 359.

   B. vii, D. 9-10 | ¹⁄₁ | V-VI, P.22, V. 1/5, A. 1 | ₁₀⁄₁₁ | , V-VI, L. l. ca. 180.

   Length of head 3½, of caudal fin 7¼, height of body 7 in the total length. Eyes—
diameter 3½ to 4½ in the length of the head, 1½ diameters from the end of the snout, and
1 apart. Snout more pointed than in S. microlepidotus. Teeth—in jaws stronger than in
the last species, and well developed on the vomer and palatines. Colours—similar to
those in the last species, with the addition of two or more rows of dark spots along the
back and also some transverse streaks.
   Habitat.—Persian Gulf to Japan.

Page 263. Add PERCIS CYLINDRICA.
Day, Proc. Zool. Soc. 1888, p. 260.

   B. vi, D. 5/21, P. 15, V. 1/5, A. 17-18, C. 15, L. l. 44, L. tr. 2½/9.

   Length of head 4, of caudal fin 5½, height of body 5¼ in the total length. Eyes—
diameter 3½ in the length of the head, 1 diameter from the end of the snout, and ¼ of a
diameter apart. The greatest width of the head equals its length, excluding the snout.
Cleft of mouth very slightly oblique: lower jaw a little the longer: the posterior
extremity of the maxilla reaches to beneath the first third of the orbit. The greatest
depth of the preorbital equals one-third of the diameter of the eye. All the opercles
entire: a well-marked spine on the opercle and another on the suboperale, no shoulder
spine. Teeth—two enlarged ones on either side, above the symphysis of the lower
jaw: fine ones on the vomer. Fins—second dorsal spine the longest, equalling
three-fourths of the diameter of the eye. Pectoral nearly as long as the head. Ventral
one-fourth longer than the head, reaching the base of the seventh anal ray. Caudal
slightly rounded. Colours—reddish-brown, with five wide and dark vertical bands,
extending from the back to the lower surface, these bands being darkest at their edges
and disappearing about the middle of the body, where there are also some dark spots. A
brown ocellus at the upper part of the base of the caudal fin, which has some brown spots
on it. Numerous brown spots on the snout and upper surface of the head and cheeks,
some on the upper edge of the eye, where there are two dark narrow bands. Ventrals
white. First dorsal fin nearly black between the spines: soft dorsal and anal with fine
dots between the rays.
   Habitat.—Two small specimens from the Andamans.

Page 264. For SILLAGO DOMINA read S. PANIJUS. Add synonym.
Cheilodipterus panijius, Ham. Buch, Fish Ganges, pp. 57, 367.

Page 267. For PSEUDOCHROMIS XANTHOCHIR read S. FUSCUS. Add synonym.
Pseudochromis fuscus and *adustus*, Müll. and Trosch. Horæ Ich. 1849, p. 23, t. iv, f. 2 ; Bleeker, Atl. Ich. ix, Sciænidæ, t. v, f. 4.

Col. Tickell figured two varieties of a species of this genus taken at Saddle Island, off Kyouk Phoo. He gave the D. 22, A. 13-14, and stated that the scales were large. The one he termed *Malucocanthus coccinicauda* being of dark burnt umber colour, becoming a little purplish below. Fins pale brown. Dorsal rays vermilion. Anal with a pale red band along its centre. Caudal deep carmine. The second, *M. bicolor*, had the anterior half of its body yellow olive-green, its posterior half superiorly including eyes, dorsal, caudal, and anal fins sepia, upper and lower margins and angle of caudal whitish gray. Pectoral and ventral yellowish. A row of small irregular spots of a smalt colour along the middle of the posterior half of the body.

Page 278. Add Genus 2—TRIGLA, Artedi.

Hoplonotus, Guichenot.

Branchiostegals seven : pseudobranchiæ present. Head parallelopiped, with its superior and lateral surfaces bony. Villiform teeth in both jaws, and usually on the vomer, but none on the palatines. Two dorsal fins, the first being of less extent than the second : three free filaments at the base of the pectoral fin. Air-bladder well developed, generally provided with lateral muscles, and sometimes partially divided internally by partitions. Pyloric appendages few or in moderate numbers.

Geographical distribution.—Coasts of Europe, and one species extending across the North Atlantic to the western shores of North America. To the south it passes round the west coast of Africa from the Atlantic to the Indian Ocean, and one species has been obtained in the Persian Gulf on one hand, and also in Japan ; consequently it is here inserted as Indian.

1. TRIGLA HEMISTICTA.
Temm. and Schlegel, Fauna Japon. Poiss. p. 36, pl. xiv, f. 3, 4, pl. xiv, B.; Günther, Catal. ii, p. 201.

Trigla arabica, Boulenger, Proc. Zool. Soc. 1887, page 663.

B. vii, D. 7/11-12, P. 11 + iii, V. 1/5, A. 11-12, C. 16.

Length of head about 3, of caudal fin 5¼, height of body 5⅓ in the total length. *Eyes*—1¼ diameters from the end of the snout, and 1½ apart. Profile from upper edge of orbit to the snout scarcely concave. Preorbital produced anteriorly into a flattened spine; two spines on the preopercle, the upper the larger. Opercle ending posteriorly in a strong spine equalling the diameter of the orbit; shoulder bone with two spines. *Teeth*—villiform. *Fins*—dorsal spines strong, the third and fourth the longest. A bony plate along the base of the dorsal fin, wider in small than in large examples. Pectoral reaches to above the third anal ray, three free appendages. *Scales*—small. *Colours*—upper part of body rosy, with numerous small rounded or oblong black spots: lower half of body white. First dorsal with a large black oblong blotch and a row of small round black dots along the middle of the second dorsal: inter-radial membrane of pectoral bluish-black.

Habitat.—An example 9 inches long has been obtained from Muscat, the species has likewise been brought from Japan.

Page 278. Genus 3—PERISTETHUS, Kaup.

Branchiostegals seven : pseudobranchiæ present. Head parallelopiped in shape, the sides and upper surface cuirassed with bone : the preorbital prolonged anteriorly into a flat projecting process. One or more barbels on the lower jaw. Teeth absent. One or two dorsal fins, the posterior of which is most developed. Two free pectoral appendages. Body covered with bony, scale-like plates. Pyloric appendages few, or in moderate numbers. Air-bladder present.

Geographical distribution.—From the south coast of Britain to the Mediterranean, also from the Atlantic and Indian Oceans to China. It has likewise been obtained at the Sandwich Islands in the South Pacific Ocean.

Page 278. PERISTETHUS HALEI.

Peristethus, Haly, The Taprobanian, vol. i, 1886, p. 165.

B. vii, D. 7/15, V. 1/5, A. 15, L. l. 34.

" Preorbital processes short, their length being contained 3½ times between their extremity and the anterior margin of the orbit. A pair of spines on the occiput, on either side of which is a low ridge terminated by a small spine. Anterior vertical plates longer than broad. The opercular ridge forms a strong spine. Lower jaw with barbels. *Colours*—uniform red."

Habitat.—A single specimen taken at Galle in deep water in April, 1883.

Page 279.   DACTYLOPTERUS ORIENTALIS. Add synonym.

Corystion orientalis, Bleeker, Waigou, 1868, p. 3.

Page 284.   Among Sir W. Elliot's and Dr. T. C. Jerdon's MS. illustrations of Indian fishes are several undescribed gobies, but as the notes respecting them have been lost, I can merely give such details as are shown on the drawings; the subject of scales and teething being omitted, must be ascertained by future observers. No specific names are attached, as the descriptions are merely for the purpose of directing the attention of collectors to the forms.

Page 284.   GOBIUS ?

Natsuli, Jerdon.

D. 6/$\frac{1}{13}$, A. 13.

Length of head 4⅔, of caudal fin 6, height of body 6 in the total length. *Eyes*—rather high up, diameter 4½ in the length of the head, 1½ diameters from the end of the snout. Cleft of mouth oblique, lower jaw somewhat the longer. *Teeth ? Fins*—spines and rays somewhat filamentous, and of about equal height, nearly equalling that of the body. Caudal rounded. *Scales ? Colours*—of a light buff, with a row of oval brown spots along the middle of the body, and several scattered smaller ones above; among these are interspersed many small yellowish-red dots. First dorsal fin with a row of orange spots along its base and a dark outer margin. Second dorsal with a similar row of orange spots along its base, a dark band along its centre, and a dark outer edge. Ventrals black. Anal with two orange bands and a dark outer edge. Caudal with 6 or 7 narrow vertical brown or orange bands, and a dark outer edge.

Habitat.—Madras, to 4 inches in length.

Page 284.   GOBIUS ?

D. 7/13, A. 13.

Length of head 6, of caudal fin 4½, height of body 7 in the total length. *Eyes*—very high up and of moderate size. *Fins*—dorsal with a short interspace, somewhat higher than the body. Pectoral longer than the head, caudal somewhat lanceolate. *Colours*—buff, becoming pink beneath, a row of cloudy spots along the middle of the sides, and indistinct bands. Numerous fine black dots on the back. A black mark under the eye. A black spot on the last two dorsal spines, both dorsal fins and upper half of caudal spotted. A dark base to the pectoral fin.

Habitat.—Adyar River near Madras, to 2·8 inches in length.

The other two forms are as follows :—No. 1, elongated, height about one-twelfth of its length. Eyes high up. Pectoral fin short. Caudal lanceolate. Buff-coloured, becoming white beneath. Fins immaculate, except the caudal which is irregularly spotted. Madras. No. 2. Height 8 in its total length. Eyes high up. Pectoral fin longer than the head. Caudal lanceolate, light brown, irregularly banded : two dark bands from the eye : a large black spot on the upper portion of the first dorsal fin : caudal irregularly spotted. In another figure a black ocellated spot may be present on the hind edge of the last dorsal rays ; while in a third the spot on the first dorsal is absent.

Page 284.   Add GOBIUS GYMNOCEPHALUS.

Bleeker, Batavia, page 473 ; Günther, Catal. iii, p. 75.

Karum natsooli, Tam.

B. v, D. 6/$\frac{1}{13}$-$\frac{1}{14}$, P. 17, V. 1/5, A. $\frac{1}{14}$, C, 13.

Length of head 6, of caudal fin 4½, height of body 8 to 9 times in the total length. *Eyes*—high up, diameter 4½ in the length of the head, ⅔ of a diameter from the end of the snout, and placed close together. Head higher than broad : snout obtuse : cleft of mouth oblique, the maxilla reaching to below the hind edge of the eye. *Teeth*—canines in both jaws. *Fins*—dorsal spines flexible, nearly as high as the body : caudal lanceolate. *Scales*—minute. *Colours*—greenish stone colour, becoming lightest beneath : three or four vertical bands on the body and another on the nape, with indistinct narrow intermediate ones. Dorsal fins darkish, unspotted : caudal also dark and reddish externally, said to have several blue and red streaks. Anal with a narrow and nearly central band along its extent, which is red externally and blue inferiorly.

Habitat.—Madras to the Malay Archipelago. Jerdon's figure is 6·4 inches in length. He has likewise the figure of another fish with much the same proportions, but the number of rays is not enumerated. He termed it *Natsi candai*, Tam. Body of a light colour, four horizontal narrow red lines along the first and three along the second dorsal fin, two along the anal which has likewise a dark outer edge. Three narrow red vertical bands down the base of the caudal fin, which has an outer dark margin.

Page 286. GOBIUS VIRIDIPUNCTATUS. Add synonymy.

Gobius chlorostigma, Bleeker, Blen. en Gob. p. 27.

Page 288. Add GOBIUS THURSTONI.

B. v, D. 6/$\frac{1}{10}$, P. 22, V. 1/5, A. 10, C. 14, L. l. 30, L. tr. 8.

Length of head 4¾, of caudal fin 4⅜, height of body 5½ in the total length. *Eyes*—upper margin near the dorsal profile, diameter 4¼ in the length of the head, 1¼ diameter from the end of the snout and 1 apart. Head ⅓ wider than long, while its height equals its length without the snout. An oblique rise from snout to eyes, from whence the dorsal profile is nearly straight: the width of the body equals ⅔ of its height. Upper jaw slightly the longer, cleft of mouth rather oblique, the posterior extremity of the maxilla hardly reaching to beneath the front edge of the eye. A single row of warts across the cheeks No barbels. *Teeth*—villiform with an outer enlarged row, and a small canine in either jaw. *Fins*—First dorsal separated by a short interspace from the base of the second dorsal, its spines flexible, the longest equalling the height of the body below it: the last rays of the second dorsal somewhat prolonged, ¼ longer than the dorsal spines, and reaching to the base of the caudal fin. Pectoral as long as the head, its upper edge straight, its lower rays the shortest, some of its upper rays silk-like. Ventral reaches vent; anal similar to second dorsal; caudal wedge-shaped. *Scales*—strongly ctenoid and angular, anterior to the dorsal fin comparatively small, there being 11 rows between the posterior edge of the orbit and the first dorsal spine : 8 rows between the bases of the second dorsal and anal, none on the head. *Colours*—slaty-grey, with 5 rows of dark and interrupted narrow brown bands in the anterior portion of the body, becoming brown spots from the pectoral fin on the base of which are two well-marked brown blotches. Numerous small blue spots on the body : first dorsal with brown spots: ventral black, and with a dark outer edge.

Habitat—One specimen 4½ inches long, sent by Mr. Thurston from Madras.

Page 291. Add GOBIUS MICROLEPIS.

Gobius acutipinnis, var. Cantor, Catal. p. 184.

,,　　microlepis, Bleeker, Verh. Bat. Gen. xxii, Blenn. en Gob. p. 35, and Java, ii, p. 436 ; Günther, Cat. iii, p. 49.

Oxyurichthys microlepis, Bleeker, En. Species, p. 120.

B. v, D. 6/$\frac{1}{13}$, P. 22, V. 1/5, A. $\frac{1}{13}$, C. 17, L. l. 50.

Length of head 6, of caudal fin 3 to 3¼, height of body from 7¼ to 9 in the total length. *Eyes*—high up, and placed rather close together ; diameter, 4 in the length of the head, and about 1 diameter from the end of the snout. Cleft of mouth oblique, lower jaw the longer, the maxilla reaches to below the hind edge of the eye. Snout obtuse. *Teeth*—in a single row without canine, those in the upper jaw a little longer and further apart than those in the lower jaw. *Fins*—both dorsals higher than the body, in some examples the fifth ray of the first dorsal fin has a filamentous prolongation. Caudal lanceolate. *Colours* —greenish or brownish-buff, with some clouded spots on the back and sides, a black dot at the edge of most of the scales : sometimes a black spot at the base of the caudal fin. Fine dark spots on the rays of the dorsal fins ; anal and caudal stained dark, especially externally.

Habitat.—Madras, to the Malay Archipelago and China.

Page 296. GOBIUS SADANUNDIO. Add synonym.

Gobius pleurostigma, Bleeker, Blenn. en Gob. p. 28.

Page 297. Add GOBIUS LITTOREUS.

Day, Proc. Zool. Soc. 1888, page 261.

B. v, D. 6/11, P. 15, V. 1/5, A. 10, C. 14, L. l. 22, L. tr. 6.

Length of head 4½, of caudal fin 4½, height of body 5½ in the total length. *Eyes*—

diameter 3 in the length of the head, $\frac{1}{2}$ a diameter from the end of the snout and placed close together. The greatest width of the head equals $\frac{4}{5}$ of its length, while its height equals its length excluding the snout. Anterior profile of head somewhat obtuse. Cleft of mouth oblique, lower jaw slightly the longer : the posterior extremity of the maxilla reaches to beneath the first third of the eye. Preopercle spineless, and no warts on the head. *Teeth*—in villiform rows, none enlarged. *Fins*—dorsal spines of moderate strength, the longest nearly half the length of the head. Pectoral as long as the head, some of its rays fine and silk-like : caudal pointed. *Scales*—finely ctenoid, none on the head : eleven rows between the occiput and front edge of the dorsal fin. *Colours*—yellowish with a few dark spots on the body and a dark band from the eye to the snout, also a dark mark on the opercle. Upper half of eye black. Dorsal, anal and caudal fins with a gray outer edging : ventrals white.

Habitat.—A small species from Madras.

Page 297.   For GOBIODON QUINQUE-STRIGATUS read G. RIVULATUS.   Add synonyms.

Gobius rivulatus, Rüppell, Atl. Fisch. p. 136, and N. W. F. p. 138.
?   „   *histrio*, Cuv. and Val. xii, p. 132, pl. cccxlvii.
Gobiodon rivulatus, Günther, F. Südsee, ii, p. 180, t. cix, f. F. and G.

Page 299.   SICYDIUM HALEI.

B. v, D 6/12, P. 19, V. 1/5, A. 11, C. 14, L. l. 56, L. tr. 16.

Length of head $5\frac{1}{2}$, of caudal fin $7\frac{1}{4}$, height of body 7 in the total length. *Eyes*—upper margin on dorsal profile, diameter $4\frac{1}{3}$ in the length of the head, $1\frac{1}{4}$ diameters from the end of the snout, and 2 apart. Body subcylindrical. Interorbital space nearly flat, snout obtuse and rounded, an oblique fall from orbit to it. Upper jaw the longer and overhung by the snout : cleft of mouth nearly horizontal : the maxilla reaches to below the middle of the eye. Lips thick. No warts, barbels or scales on the head. *Teeth*—in maxilla, in a single external movable row in the gums, directed almost horizontally, and a single inner row of longer pointed and curved ones, these two rows being divided by a considerable interspace : a large recurved canine on either side of symphysis of the lower jaw : in a single row of much smaller teeth in the upper jaw. *Fins*—spines of first dorsal ending in filamentous prolongations, but the longest is not quite so high as the body below it. A considerable interspace between the first and second dorsal fins, the rays of the latter are equal to about half the height of the body, and similar to the anal. Ventral does not extend half way to the anus. Caudal rounded at the extremity, its central rays somewhat the longest. Pectoral as long as the head, excluding the snout. *Scales*—strongly ctenoid, the exposed portion above twice as high as wide, and rounded, about 19 rows from occiput to first dorsal fin, the first few anterior rows somewhat small, the remainder on the body of about the same size. *Colours*—greenish brown, a black interorbital band which is continued from the eye to the angle of the mouth: some dark vertical bands on the body : a dark outer edge to ventral and anal, also a dark band to outer edge of caudal, margined externally with white, which is widest at the angles.

Habitat—Ceylon, from whence Mr. Haly has sent me an example 3 inches long.

Page 310.   ELEOTRIS MACROLEPIDOTA.

This fish is not *E. hoedtii*, &c. Bleeker, as observed in Günther's "Fische der Südsee," ii, p. 185, as the type at Berlin (No. 2155) has D. 7/$\frac{1}{3}$, A. $\frac{1}{10}$, the last ray in both being almost double, and therefore counted as two by Bloch. L. l. 30, L. tr. 13-14, and from 26 to 28 scales between the snout and first dorsal fin.

Page 310.   ELEOTRIS MURALIS.   Add synonym.

Eleotris lineato-oculatus, Kner, SB. Wien. Ak. lvi, p. 720, t. iii, f. 1.

Page 311.   Add ELEOTRIS ELLIOTI.

Day, Proc. Zool. Soc. 1888, p. 262.
Cul nachooli, Tamil.

B. vi, D. 6/12, P. 21, V. 6, A. 13, C. 13, L. l. 80, L. tr. 16.

Length of head $4\frac{1}{2}$, of caudal fin $4\frac{2}{3}$, height of body $5\frac{1}{2}$ in the total length. *Eyes*—high up, diameter $3\frac{1}{2}$ in the length of the head, 1 diameter from the end of the snout. Height of head $\frac{3}{4}$ of its length: interorbital space narrow. Cleft of mouth somewhat oblique, the maxilla extends posteriorly to beneath the middle of the eye. *Teeth*—rather large, in single row in the upper jaw with two small lateral canines : in two or three rows in the centre of the lower jaw, separated from the single lateral row by two large recurved canines.

Fins—dorsal spines thin, flexible and equal in height to the body below them, second dorsal and anal of similar height and one-third lower than the first dorsal. Pectoral nearly as long as the head. Caudal rounded with its central rays somewhat the longest. *Scales*—ctenoid in the posterior portion of the body, where they are larger than in the anterior portion, and small on the surface of the head : none on the cheeks. *Colours*— whitish with five wide and light chestnut bands descending from the back, each of which has a black outer edge : another over the nape is without dark edges. Caudal fin brown, with a broad yellowish black-bordered vertical band down its centre. A dark horizontal band running along the cheeks below the eye. Dorsal fins light brown with white outer edges, a large black white-edged blotch in the posterior half of the first dorsal fin, and a second but smaller one at the termination of the second dorsal, which last fin is white at its base.

Habitat.—Madras. A skin from Sir W. Elliot's collection is 3·2 inches in length, but it is in a bad condition. A coloured drawing was made when the fish was fresh.

Page 312. ELEOTRIS POROCEPHALUS. Add synonyms.

Eleotris ophiocephalus, Cuv. and Val. xii, p. 239 ; Günther, Fische Südsee, ii, p. 185, t. cxii, f. A.
Eleotris viridis, Bleeker, Madura, p. 22.
Ophiocara ophiocephala, Bleeker, Eleotriformes, 1874, p. 15.

Page 312. For ELEOTRIS OPHIOCEPHALUS read E. TUMIFRONS. Add synonyms.

Eleotris tumifrons, Cuv. and Val. xii, p. 241.
Ophiocara koedtii (young), tolsoni (young), and *aporos*, Bleeker, Eleotriformes, 1875, pp. 33, 35.
Eleotris macrolepidotus, Günther, Fische, Südsee, ii, p. 186 (not Bloch).
Eleotris macrocephalus, Günther, l. c. t. cxii, f. B.

Page 323. Add

FAMILY—TRICHONOTIDÆ, *Günther*.

Branchiostegals seven : pseudobranchiæ. Gill-openings wide. Body elongated, sub-cylindrical. The infraorbital ring of bones does not articulate with the preopercle. Teeth mostly villiform. One or two dorsal fins occupying almost the entire length of the back, when there are two, the first is short and the anal similar to the second dorsal. Fin rays branched. Ventrals jugular with one spine and five rays. No prominent papilla near the vent. Scales cycloid of moderate size. Air-bladder and pyloric appendages absent.

The fishes of this family have been variously located. A species of HEMEROCŒTES was placed by Forster and also by Schneider among the *Callionymidæ*, and near which Cuv. and Val. considered it should be located. Dr. Günther (Catal. Fishes Brit. Museum, ii, p. 225) observed that it "is not an Acanthopterygian fish, all its fin rays being articulated." Subsequently he remarked (l. c. iii, 1861, p. 484), that the affinities of these fishes are very obscure, and instituted an Acanthopterygian family for their reception, observing that the ventral fin had one spine and five rays, he placed it between the Ophiocephalidæ and Cepolidæ, and in 1880 he located it among the Acanthopterygii Blenniiformes. Steindachner, in 1867, suggested that a species he described might possibly be a type of labroids, but the example was too small to examine the pharyngeal bones.

Geographical distribution.—Small fishes of the seas and coasts of India, and the Malay Archipelago to New Zealand.

Genus V.—TRICHONOTUS, *Bl. Schn.*

Head depressed and pointed, with the lower jaw the longer. Cleft of mouth deep, almost horizontal, the lower jaw the longer. Eyes of moderate size, closely approximating. Conical teeth in jaws, vomer, and palatine bones. One long dorsal fin, the first few rays may be elongated, or else slightly detached.

Habitat.—Andamans to the Malay Archipelago.

1. TRICHONOTUS SETIGERUS.

Bl. Schn. p. 179, t. xxxix ; Cuv. and Val. xii, p. 316 ; Bleeker, Celebes, v, p. 251 ; Günther, Catal. iv, p. 484.

Trichonotus polyophthalmus, Bleeker, Ceram. iii, p. 243 (*female*).

B. vii, D. ⁴/₅, P. 11, V. 1/5, A. 37, C. 13, L.l. 58, L. tr. 6.

Length of head 4, of caudal fin 6¼, height of body 10 in the total length. *Eyes*—diameter ⅓ of the length of the head, 1 diameter from the end of the snout, and placed close together, so that they are directed somewhat upwards. *Teeth*—a single row in the jaws, vomer and palatines, being somewhat enlarged in the intermaxillaries. *Fins*—owing to the small size of the example, it is difficult to count the number of rays. The dorsal commences above the axil of the pectoral, its two first rays are not elongated (? age or sex), but slightly divided from the remainder of the fin.

Habitat.—This example, measuring a little over 1¼ inches in length, was obtained at the Andamans.

Page 324.   Add CEPOLA INDICA.

B. vi, D. plus quam 90, P. 17.   A. plus quam 90.

Length of head 8, height of body 8 in the total length. *Eyes*—diameter 3¼ in the length of the head, ⅓ a diameter from the end of the snout, and ⅔ of a diameter apart. Cleft of mouth oblique, the maxilla reaches posteriorly to beneath the middle of the eye. A strong spine at the angle of the preopercle, one on the vertical limb above it, and four on the horizontal limb. *Teeth*—in a single row in both jaws, a small curved canine in an outer row in the lower jaw, also one in upper but not in a separate row. *Fins*—the dorsal commences on a line slightly posterior to the orbit, its rays are unbranched, they increase in height to the sixth, which is 2/3 that of the body below it, from whence they gradually decrease and join with the caudal, there appear to be over 100 rays. Anal begins beneath the ninth dorsal spine, and has nearly as many rays as the dorsal, it is conjoined to the caudal, the latter being pointed. *Scales*—small but distinct, they appear as if forming horizontal ridges, cheeks scaled, none on the opercles. *Lateral-line*—commences from above the middle of the upper margin of the opercle, then ascends to close to dorsal fin and becomes obsolete after first third of the body. *Colours*—of brick-dust red, dorsal and anal fins with dark outer edges, an oval black spot between eighth and eleventh dorsal rays.

Habitat.—Madras, from whence Mr. Thurston has sent me one specimen 8 inches long.

Page 325.   For BLENNIUS LEOPARDUS read SALARIAS BREVIS.   Add synonym.

Salarias brevis, Kner, SB. Wien Ak. lviii, 1866, p. 334, t. vi, f. 18; Günther, Fische Südsee, ii, p. 203, t. cxviii, f. c.

Page 326.   For BLENNIUS STEINDACHNERI read SALARIAS STEINDACHNERI.

Page 327.   PETROSCIRTES VARIABILIS.   Add synonym.

? *Petroscirtes petersi*, Koss. and Raüb, F. R. M. p. 21, t. ii, f. 9.

Page 328.   Add PETROSCIRTES STRIATUS.

Day, Proc. Zool. Soc. 1888, p. 262.

B. vi, D. 40, P. 13, V. 3, A. 27, C. 10.

Length of head 4¾, of caudal fin 6⅔, height of body 6 in the total length. *Eyes*—diameter 2⅔ in the length of the head, ⅔ of a diameter from the end of the snout, and the same distance apart. The greatest width of the head equals half its length: the maxilla reaches to below the first third of the orbit. Snout somewhat broad and rounded in front, the upper jaw a little the longer. No tentacles on the head. *Teeth*—an exceedingly large recurved canine on either side of the lower jaw, and a much smaller one in the upper, while about 14 teeth exist in a single row in each jaw between the canines. *Fins*—dorsal commences midway between the eye and hind edge of the opercles, and does not extend quite so far as the caudal fin, the height of its longest rays equals two-thirds of that of the body, and rather more than those in the anal fin, which latter is not united to the caudal. *Colours*—with about ten broad vertical bands extending from the base of the dorsal to the anal fins, separated from one another by a very narrow light line.

Dorsal and anal fins externally black edged, and the membrane studded with fine brown spots. Caudal light-coloured.

Habitat.—Ceylon, one specimen 1½ inches in length.

Page 330. SALARIAS FUSCUS. Add synonym.

Salarias phaiosoma, Bleeker, Batoe, p. 317.
    „   *holomelas*, Günther, Ann. and Mag. Nat. Hist. x, 1872, p. 399.
    „   *niger*, Koss. u. Raüb. F. R. M. p. 21, t. ii, f. 8.

Page 331. Add SALARIAS SINDENSIS.

Day, Proc. Zool. Soc. 1888, p. 263.

B. vi, D. 13/20, P. 14, V. 2, A. 23, C. 12.

Length of head 1/5, height of body 1/5 of the total length. *Eyes*—situated high up near the dorsal profile, diameter 1/4 of the length of the head, 1 diameter from the end of the snout and also apart. Body strongly compressed, profile from above the orbits to the end of the snout oblique. The height of the head equals its length excluding the snout. The posterior extremity of the maxilla reaches to beneath the front edge of the eye. No tentacles or crest on the head. *Teeth*—large, well developed, posterior canines. *Fins*—dorsal not notched, but becoming higher posteriorly where the longest rays equal half the height of the body: anal not quite so high as soft dorsal: dorsal, anal, and caudal rays unbranched. The dorsal and anal fins not quite connected to the caudal. *Colours*—olivaceous, four wide brown bands on the head, the three anterior of which encircle it, about twelve vertical bands on the body, more or less distinct, but most so at the base of the dorsal fin. Dorsal fin with a dark mark along its anterior two-thirds: anal black-edged, each ray tipped with pure white. In one there appears to be marks of some narrow, horizontal bands having existed along the front half of the body.

Habitat.—Three specimens up to 2½ inches in length from Kurrachee in Sind.

Page 331. Add SALARIAS CRUENTIPINNIS.

Tickell, Fishes, p. 313, MSS. with a figure.

B. vi, D. 13/13, V. 2, A. 17.

Length of head 5, of caudal fin 5½, height of body 4 in the total length. *Eyes*—high up near the dorsal profile. Body compressed: the profile from the eyes to the mouth almost vertical: the posterior extremity of the maxilla reaches to beneath the hind edge of the eye. No crest on the head: a bifurcated supraorbital tentacle and a fringed nasal one. *Fins*—dorsal not notched, and posteriorly continued on to the caudal fin, its spinous portion equal to three-fourths the height of the body and rather more than its soft part. Anal lower than the dorsal, its posterior rays the longest. *Colours*—rich vinous olive sepia: a large patch of pale yellowish-brown from the angle of the lips to the lower edge of the subopercle. Caudal fin of the same colour as the body, with the three outer rays above and below tawny. Dorsal dusky, its basal half blackish, external half of anterior 17 rays carmine, of the 9 posterior rays black. Anal fin dusky with a carmine band along its centre, and externally with a carmine and black edging. Pectoral paler than the body with its lower rays tinged with carmine.

Habitat.—Saddle Island, off Kyoukphoo in Arracan. The specimen was 2·8 inches in length.

Page 332. Add SALARIAS NEILLI.

Day, Proc. Zool. Soc. 1888, p. 263.

B. vi, D. 12/17, P. 13, V. 2, A. 19, C. 10.

Length of head 4½, height of body 4½ of the total length. *Eyes*—situated high up near the dorsal profile, 4 diameters in the length of the head, 1 diameter from the end of the snout and half a diameter apart. Frontal profile very steep, the head as high as it is long, the maxilla reaches to beneath the last third of the eye. A fringed supraorbital tentacle about twice as long as the eye, a small fringed one at the nostril, no crest on the head. *Teeth*—in a single row fixed, a very large curved canine posteriorly in the lower jaw and a smaller curved one in the upper. *Fins*—spinous portion of dorsal fin lower than the rayed part, the notch between the two portions well marked, the longest dorsal rays are equal to half the height of the body of the fish, neither the dorsal nor anal fins are attached to the caudal, which latter is somewhat wedge-shaped and its rays are branched. *Colours*—olive with seven or eight short dark bands descending from the dorsal fin down the first third of the body. Some dark bands radiate from the eye: a large black blotch below and somewhat behind the orbit. Two semicircular brown bands across the lower surface of the mandibles. Fins darker than the body.

I have named this fish after A. Brisbane Neill, Esq., to whom I am under great obligations for the valuable assistance he has always afforded me in my publications.

Habitat.—Kurrachee in Sind, out of ten specimens the longest is 2¼ inches.

Page 332. SALARIAS LINEATUS. Add synonym.

Salarias caudolineatus, Günther, Fische Südsee, ii, p. 209, t. cxvi, f. F.

Page 333. Add SALARIAS OORTII.

Bleeker, Nat. Tyds. Ned. Ind. i, p. 257, f. 15, and Act. Soc. &c. Indo-Ned. iii, Sumatra, p. 39 ; Günther, Catal. iii, p. 257.

B. vi, D. 12/19-21, P. 14, V. 2, A. 23-24, C. 13.

Length of head 7, of caudal fin 7, height of body 7 to 7½ in the total length. *Eyes*—high up, diameter ¼ of the length of the head, 1¼ diameters from end of snout, which is very slightly oblique. The maxilla extends to somewhat beyond the hind edge of the orbit. A crest on the summit of the head, a fringed tentacle above the orbit and another at the nostrils. *Teeth*—small canines in the lower jaw. *Fins*—dorsal fin deeply notched almost to its base, while posteriorly it is continuous with the caudal : its anterior portion two-thirds as high as the body, and its posterior at least one-third higher : caudal rounded, its central rays being the longest. *Colours*—stone-colour along the back, becoming violet on the side and beneath : darker bands from the back, sometimes arranged in pairs. Anterior dorsal reddish-violet, with several undulating narrow white lines and sometimes a small black blotch between the first and second spine. Second dorsal with the white bands taking an oblique direction upwards and backwards : bluish marks or spots in its outer fourth. Caudal and anal with their outer thirds brownish.

Habitat.—Aden, the east coast of India to the Malay Archipelago.

Page 334. SALARIAS ALBOGUTTATUS. Add

Kner, SB. Wien Ak. lvi, 1867, f. 6 ; Günther, F. Südsee, ii, p. 205, t. cxviii, f. B.

Page 335. SALARIAS MARMORATUS. Add synonym.

Salarias arenatus, Bleeker, Cocos. iii, p. 173 ; Günther, Catal. iii, p. 249.

Page 335. Add SALARIAS BICOLOR.

Salarias bicolor, Tickell, MSS. with a figure.

D. 11/17, V. 2, A. 18.

Length of head 5, of caudal fin 5, height of body 6 in the total length as shown by the figure, snout not overhanging the mouth, no crest on head, tentacles were not observed. *Fins*—first dorsal as high as the body below it and separated by a deep notch from the second dorsal, which last is not confluent with the caudal. *Colours*—anterior half of the body so far as to the origin of second dorsal fin of a deep blue (smalt), posterior half carmine-orange. Dorsal fins sepia tinged with smalt, base of second dorsal orange : caudal and anal orange tipped and margined with sepia : pectoral smalt : ventrals whitish.

Habitat.—A specimen 1⅜ inches long from Saddle Island, Kyoukphyoo, Arracan.

Page 336. Add—Genus ACANTHOCLINUS, *Jenyns.*

Six branchiostegals : pseudobranchiæ. Body elongate. Cleft of mouth of moderate width. Gills united beneath the throat. Teeth in jaws, vomer and palate. Dorsal fin single, occupying most of the length of the back, it is chiefly composed of spines : anal long, and having more spines than rays. Ventral jugular consisting of one spine and three rays : caudal distinct. Scales cycloid : lateral-line present or absent. No air-bladder.

Geographical distribution.—Coasts of India and New Zealand.

1. Acanthoclinus indicus.

Day, Proc. Zool. Soc. 1888, p. 264.

B. vi, D. 21/4, P. 16, V. 1/3, A. 10/4, C. 17, L. l. 40, L. tr. 14.

Length of head 4, of caudal fin 5, height of body 3 in the total length. *Eyes*—diameter 1/5 of the length of the head, 1 diameter from the end of the snout, and ¾ of a diameter apart. Cleft of mouth somewhat oblique, the maxilla reaching posteriorly to beneath the hind third of the orbit. Two strong opercular spines. *Teeth*—in jaws, vomer, and palate. *Fins*—dorsal spines strong, the fins not united with the caudal: pectorals rounded: ventrals long and inserted slightly in front of the base of the pectoral: caudal rounded. *Scales*—cycloid. *Lateral-line*—absent. *Colours*—brownish-black with a milk-white band commencing on the front end of the dorsal fin, and extending to the snout: a white band over the free portion of the tail: a white spot at the base of the pectoral fin: one on either side of the base of the mandibles, one on the isthmus. The posterior half of the ventral fin, also a ring round the vent, white: as well as the tip of the caudal fin.

Habitat.—Madras, where one example, an inch long, was captured.

Page 336. Add Genus—CRISTICEPS, *Cuv. and Val.*

Branchiostegals six: pseudobranchiæ. Body elongate covered with small or rudimentary scales. Gill-opening wide. Cleft of mouth of moderate width, snout short. Usually some tentacles on the head. Fine teeth on the jaws and vomer. Two separate dorsal fins, the anterior being composed of three spines, the posterior with many rays, the majority of which are spines. Ventral jugular with one spine and two or three rays. Pyloric appendages absent. Viviparous.

Habitat.—Mediterranean, Ceylon to the Malay Archipelago, coasts and rivers of Australia and Tasmania.

Cristiceps halei.

B. vi, D. 3/Ψ, P. 13, V. $\frac{1}{2}$, A. $\frac{1}{5}$, C. 14.

Length of head 4½, of caudal fin 7, height of body 4½ in the total length. *Eyes*—4 diameters in the length of the head, 1 diameter from the end of the snout and nearly 1 apart. A broad-fringed supraorbital tentacle and a short simple one on the snout. *Teeth*—fine in the jaws and on the vomer, none on the palatines or tongue. *Fins*—first dorsal commences above a vertical line from the hind edge of the eye, and its spines are higher than the front ones in the second dorsal fin, it is not confluent with the caudal. All the pectoral rays unbranched, anal commences below about the eighth spine of the second dorsal: anal with two spines fifteen unbranched and four divided rays. Caudal wedge-shaped. *Scales*—rudimentary. *Lateral-line*—with a rather strong curve anteriorly. *Colours*—brownish-yellow with a white mark behind the lower half of the orbit, and some irregularly-shaped similar markings on the occiput and gill-covers, two more at the base of the pectoral fin, a row of about 12 below the base of the spinous dorsal fin, and two more badly developed rows along the sides of the body, of which the three largest are behind the pectoral fin and are longer than wide.

Habitat.—Colombo, where it was obtained by Mr. Haly, who is doing such good work among the Ceylon fishes and after whom I have named the single specimen obtained, and which is figured life size.

Page 337. XIPHASIA SETIFER. Add synonyms.

? *Nemophis lessonii*, Kaup, Proc. Zool. Soc. 1858, p. 168.
? *Xiphogadus madagascarensis*, Playfair, P. Z. S. 1868, p. 11.
Xiphasia setifer, Ramsay and Ogilby, Linn. Soc. N. S. W. i, 1886, p. 582.

B. vi, D. 128-129 (233 ?), P. 13, V. 3, A. 115-116, C. 12.

Length of head 16, of caudal fin 32 in the total length. *Eyes*—3½ in the length of the head, from ¼ to ⅓ of a diameter apart and 1 diameter from the end of the snout: upper profile of the head rounded. The upper jaw slightly the longer. *Teeth*—a single row of closely set, recurved, cardiform teeth in the lower jaw, and with a large lateral canine on either side, which is received into a groove in the roof of the mouth. Teeth in the upper jaw similar in size and number to those in the lower, except that the lateral canines, although present, are merely half the size of those in the mandibles. *Fins*—the dorsal commences above or slightly before the orbits, and extends posteriorly to the root of the caudal fin to which it is not joined: the anal begins beneath the seventeenth dorsal ray and similarly reaches the root of the caudal fin. The rays of both fins are simple, unbranched, and higher than the body. In the Australian examples the caudal fin was distinct with no elongated central ray, but this last was observed by Jerdon at Madras. *Colours*—alternate bands of dark and light ash: the fins opaline: the dorsal with a black and narrow white-edged margin, becoming widened anteriorly into blotches.

Habitat.—Coromandel coast of India, and New South Wales, possibly Madagascar, and probably the South Sea. It attains at least 14 feet in length. Jerdon observed, "said to be venomous."

Page 349.   For MUGIL CARINATUS read M. KLUNZINGERI.   Omit synonym

Mugil carinatus, C. V.
Add *Mugil klunzingeri*, Day, Proc. Zool. Soc. 1888, p. 264.

Page 349.   Add MUGIL CARINATUS.
(Ehr.) Cuv. and Val. xi, p. 148.

D. 4/⅓, P. 14, V. 1/5, A. ⅔, C. 15, L. l. 38, L. tr. 12-13.

Length of head from 4¼ to 4½, of caudal fin 4¼, height of body 4¼ in the total length. *Eyes*—diameter ⅓ of the length of the head, nearly 1 diameter from the end of the snout and 1½ diameters apart. The greatest width of the head equals its length behind the last third of the eye. Eye with a narrow posterior adipose lid. Interorbital space flat. Upper lip rather thick: preorbital scaleless, moderately curved and serrated: the end of the maxilla visible. The mandibular bones form an obtuse angle: the uncurved space on the chin is broadly lanceolate. About 25 rows of scales between the snout and the origin of the dorsal fin. *Fins*—first dorsal higher than the second, its spines of moderate strength, the height of the first being equal to the width of the head: the fin commences above the tenth scale of the lateral-line, the second dorsal above the twenty-first: the pectoral reaches the eleventh scale. Soft dorsal and anal fins scaled, the latter commencing very slightly in advance of the vertical of the former. The lowest depth of the free portion of the tail equal to 2¼ in the length of the head. *Scales*—no elongated one in the axil, one along the base of the first dorsal, also at the ventral: the scales on the back from in front of the first dorsal fin form a sort of keel for some little distance. *Colours*—golden around the eye, no black pectoral spot.

Habitat.—Red Sea and seas of India.

Page 350.   For MUGIL PLANICEPS read M. TADE.   Add synonym.

Mugil tade, Forsk. p. 74; Cuv. and Val. xi, p. 153; Klunz. F. R. M. p. 133, t. x, f. 3 and 3a.

Page 353.   MUGIL CUR.   Add synonym.

Myxus superficialis, Klunz. F. R. M. synopsis. i, p. 831 *(young)*.

Page 355.   MUGIL CRENILABRIS.   Add synonyms.

Mugil cirrhostomus, Forster, Desc. Anim. pp. 198, 257.
    „   *fasciatus*, Cuv. and Val. xi, p. 125.
    „   *macrochilus*, Bleeker, 1854, p. 53.
    „   *rüppellii*, Günther, Catal. iii, p. 458.

Page 376.   REGALECUS RUSSELLII.   Add synonym.

Regalecus pacificus, Haast, Trans. N. Z. Inst. xi, p. 269.

Page 378.   AMPHIPRION SEBÆ.   Add synonym.

Prochilus sebæ, Bleeker, Nat. Verh. Holl. 1877, p. 30, and Atl. Ich. t. cccc, Pom. t. i, f. 9.

Page 379.   AMPHIPRION BIFASCIATA.   Add synonym.

Amphiprion trifasciatum, Cuv. and Val. v, p. 395.
    „   *intermedius*, Schleg. Overs Amph. &c. p. 19.
Coracinus vittatus, Gronov. ed. Gray, p. 85.
Prochilus bifasciatus, Bleeker, Nat. Verh. Holl. 1877, p. 31, and Atl. Ich. Pom. t. i, f. 4, 5, 6.

Page 381.   TETRADRACHMUM MARGINATUM.   Add synonyms.

Heliastes reticulatus, Richards. Ich. China, p. 254.
Pomacentrus unifasciatus, Kner, Sitz. Wien. Ak. 1868, lviii, p. 348, f. 24.

Page 381. Add TETRADRACHMUM TRIMACULATUM.

Pomacentrus trimaculatus, Rüpp. Atl. Fische, p. 39, t. viii, f. 3.
       „   *nuchalis*, Benn. Life of Sir S. Raffles, p. 688.
Dascyllus trimaculatus, Cuv. and Val. v, p. 441; Günther, Catal. iv, p. 13; Klunz. F. R. M. 1871, p. 519.
Dascyllus unicolor, Benn. Proc. Zool. Soc. 1831, i, p. 127.
       „   *niger*, Bleeker, Verh. Bat. Gen. xxi, Labr. &c. p. 10.
Sparus nigricans, pt. Gronov. ed. Gray, p. 61.
Tetradrachmum trimaculatum, Bleek, Atl. Ich. ix, Poma. t. x. f. 8.

B. v, D. $\frac{2}{13}\cdot\frac{1}{15}$, P. 17 V. 1/5, A. $\frac{2}{13}\cdot\frac{1}{14}$, C. 15 L. l. 27, L. tr. 3/11, Cœc. pyl. 3, Vert 11/14. Length of head 4 to 4½, of caudal fin 5, height of body a little over half of the total length. *Eyes*—diameter 2/5 of the length of the head, half a diameter from the end of the snout. Preopercle rather coarsely serrated. *Scales*—lateral-line ceases below the soft dorsal fin, but is continued in the middle of the free portion of the tail, as one or two holes in each scale. *Colours*—deep brown, vertical fins dark, becoming black at their edges. A white spot at the nape, which is sometimes wanting, a second above the lateral-line below the middle of the dorsal fin.

Habitat.—Red Sea, and east coast of Africa to Polynesia. In Sir Emerson Tennent's account of Ceylon, Dr. Günther gave this species as existing there, which has been confirmed by Haly ('Taprobanian, i, 1886, p. 166) who states it to be common at Colombo.

Page 382. POMACENTRUS TRILINEATUS. Add synonym.

Pomacentrus tripunctatus, emarginatus, vanicolensis and *chrysurus*, Cuv. and Val. v, pp. 421, 422, 423.
Pristotis fuscus, Bleeker, Bali, p. 9.
Pomacentrus tæniops, Less. Voy. Coq. Poiss. p. 189, t. xviii, f. 1.
       „   *katuako, tæniometopon* and *simsiang*, Bleeker, Timor, p. 169, Amboina and Ceram. p. 283, and Nat. Tyds. Ned. Ind. 1856, xi, p. 90.
Pomacentrus bilineatus, Castleman, P. Z. S. Victoria, ii, p. 89.

Page 384. For POMACENTRUS ALBOFASCIATUS read P. PROSOPOTÆNIA.

Omit synonyms, and insert
Pomacentrus prosopotænia, Bleeker, Singapore, p. 67.

Page 384. For POMACENTRUS PUNCTATUS read P. *lividus*. Add synonyms.

Chætodon lividus, Forsk. Desc. Anim. p. 227.
Eupomacentrus lividus, Bleeker, Atl. Ich. Pomac. t. iv, f. 5.

Page 386. Add GLYPHIDODON MELAS.

Cuv. and Val. v, p. 472; Bleeker, Verh. Bat. Gen. xxi, Labr. Cte. p. 23; Schlegel, Ov. Amph. &c. Verh. Nat. Gen. Ned. Overz. Bez. p. 23, pl. v, f. 2; Günther, Catal. iv, p. 45; Playfair and Günther, Fish. Zanz. p. 83.
Glyphidodon ater, Cuv. and Val. v, p. 473.
Paraglyphidodon melas, Bleeker, Atl. Ich. ix, t. cccciv, f. 4.
Nga yanga ap'hyoo, Arracan.

B. v, D. 13/13-14, P. 17, V. 1/5, A. $\frac{2}{16}\cdot\frac{1}{14}$, C. 17, L. l. 28, L. tr. 3/10. Length of head 4, of caudal fin 5, height of body 2¼ in the total length. *Eyes*—diameter 3¼ in the length of the head, 1 diameter from the end of the snout. The depth of the anterior portion of the suborbital ring of bones equals that of the preorbital. *Teeth*—narrow compressed. *Fins*—dorsal spines rather short, increasing in length posteriorly, the soft portion of dorsal and anal somewhat rounded, caudal slightly emarginate. *Lateral-line*—ceases below hind edge of dorsal spines. *Colours*—neutral sepia or dusky, with a greenish tinge beneath: fins black or a little diluted at their bases. Scales edged darker.

Habitat.—Red Sea, east coast of Africa, Burma to the Malay Archipelago.

Page 387. GLYPHIDODON ANTJERIUS. Add synonyms.

Glyphisodon leucopoma, Cuv. and Val. v, p. 480.
       „   *xanthozona* and *phaiosoma*, Bleeker, Sumatra ii, p. 283 and Verh. Bat. Gen. xxii, Bali. p. 9.
Glyphidodon dispar, Günther, Catal. iv, p. 53.
       „   *cingulus, albovinctus* and *henimelas*, Kner, Sitz. Wien. Ak. 1867, lvi, p. 725, lviii, p. 351, xviii, p. 351, f. 25.
Glyphidodon zonatus, immaculatus, modestus and *cyaneus*, Bleeker, Atl. Ich. Pomac. t. x, f. 2.
Glyphidodontops antjerius, Bleeker, Atl. Ich. Pomac. t. xi, f. 2.

Page 391. Add Genus—*Xiphochilus*, Bleeker.

> *Branchiostegals six : pseudobranchiæ present. Body oblong : head scaled and nearly as high as long : snout obtuse, upper lip thin and can be almost hidden under the preorbital. Both limbs of the preopercle are destitute of scales. Four canine teeth anteriorly in both jaws, while the lateral teeth are soldered into an osseous ridge : a posterior canine tooth present. Fins having the following numbers of rays, D.¹⁄¹·⁻¹ⁿ, A. ₇⁄₁₀. Scales large, 28 or 29 along the lateral-line. No scales along the bases of the fins. Lateral-line continuous.*

XIPHOCHILUS ROBUSTUS.

Günther, Catal. iv, p. 98 ; Klunz. F. R. M. 1871, p. 110.

B. vi, D. ¹⁄₉·⁻¹·¹ A. ₇⁄₁₀, L.l. 29, L. tr. 3/9.

The following is from Dr. Günther's description :—Height of body nearly 3½, length of head; 3½ in the total length. Head nearly as high as long : snout obtuse. Preorbital higher than the orbit, preopercle not serrated. *Teeth*—four strong canines in either jaw, the outer ones of the mandibles being turned outwards, an obtuse osseous ridge round the edge of the jaws in which teeth are scarcely distinct. *Fins*—Dorsal spines strong, the last being the longest and equalling one-third the length of the head, the soft dorsal and anal reach the root of the caudal, the last being rounded. *Colours*—yellowish red, a yellow band along the basal half of the anal and middle of the dorsal fin.

Habitat.—One specimen 12 inches long, obtained in Ceylon by Mr. Haly (Taprobanian, i, p. 165), and one in the British Museum is from the Mauritius : also Red Sea.

Page 392. Add COSSYPHUS BILUNULATUS.

Labrus bilunulatus, Lacép. iii, pp. 454, 526, pl. xxxi.

Cossyphus bilunulatus, Cuv. and Val. xiii, p. 121 ; Bleeker, Amboina, ix, p. 4, and Atl. Ich. i, p. 101, t. xxxviii, f. 3 ; Günther, Catal. iv, p. 105.

B. vi, D. ¹²⁄₁₀, P. 16, V. 1/5, A. ₁₃⁄₃, C. 14, L. l. 34.

Length of head 3⅔, of caudal fin about 6¼, height of body 3¼ in the total length. *Eyes*—diameter 5½ in the length of the head, and 2 diameters from the end of the snout. Preopercle finely serrated, and scaled. *Fins*—caudal emarginate, the outer rays being produced. *Colours*—reddish with light or yellow stripes and a large black blotch below the hind edge of the soft dorsal fin and over the commencement of the free portion of the tail. Two black lines on the head, one from the snout through the eye, the second from the angle of the mouth to the subopercle. A black blotch between the first three dorsal spines. *Habitat.*—Isle de France, Ceylon (Haly) to the Malay Archipelago. This fish is considered by some to be identical with *C. microrus*, Lacép. *C. chabrolii*, Lesson, *C. maldat*, Cuv. and Val. and *Labrus spilonotus*, Bennett.

Page 394. Add CHEILINUS UNDULATUS.

Rüpp. N.W. Fische, p. 20, t. vi, f. 2 ; Cuv. and Val. xiv, p. 108 ; Bleeker, Atl. Ich. i. p. 68, Labroidei, t. xxvi, f. 3 ; Günther, Catal. iv, p. 129 ; Klunz. F. R. M. 1871, p. 112. *Crassilabrus undulatus*, Swainson, Fish, ii, p. 225.

B. v, D. ⁹⁄₁₀, P. 12, V. 1/5, A. ³⁄₈, C. 11, L. l. 22-23, Vert. 9/14.

Length of head 3⅓, of caudal fin 5½, height of body about 3 in the total length. *Eyes*—diameter 1/6 of the length of the head and situated in about the middle of its length. Head slightly longer than high, and having a hump in some old specimens. Lower jaw slightly the longer : lips thick. *Fins*—ventrals not quite so long as pectorals : caudal rounded. *Scales*—two or three rows of scales on the cheeks. *Lateral-line*—tubes not branched. *Colours*—Bluish green, with the anterior half of the body below the lateral-line reddish, as are also the cheeks. Two narrow dark bands pass from the eye to the snout, between which is a yellow one : two similar bands pass backwards from the eye. Many narrow red and yellow lines on the head and chest, and dark undulating bands on the fins, outer edge of caudal yellow.

Habitat.—Red Sea, Zanzibar, Ceylon (Haly) to the Malay Archipelago.

Page 398. Add PLATYGLOSSUS METAGER.

Julis metager, Tickell, Fish. MSS. p. 322, c. fig.

B. vi, D. ₉⁄₁₁, V. 1/5, A. ₁₂⁄₃.

Length of head 4⅓, of caudal fin 7, height of body 3¼ in the total length, according to the figure. *Eyes*—diameter 4½ in the length of the head, 1¾ diameters from the end of the snout. Body compressed, the form of the dorsal and abdominal profiles about equally convex. *Teeth*—the posterior canine said to be large. *Fins*—dorsal moderately elevated, .

equal in its highest portion to about one-third the height of the body, and similar to the anal. Caudal rounded. *Colours*—body and fins of a deep olive bistre, the body longitudinally striated with about thirteen lines of sepia. An elongated patch of a black colour, and having a grayish white margin filling up the middle half of the dorsal fin from the eleventh to the fifteenth ray. Caudal tawny with a wide central and vertical brown band. *Young.*—Caudal tawny white with the band of pale Indian red.

Habitat.—The larger example which is figured, is a little over 4 inches in length; both were taken on November 27th, 1862, at Saddle Island off Kyoukphyoo.

Page 400. Add PLATYGLOSSUS JAVANICUS.

Julis javanicus, Bleeker, Java iv, p. 341.
Halichœres javanicus, Bleeker, Atl. Ich. i, p. 125, Labroidei, pl. xl, f. 3.
Platyglossus javanicus, Günther, Catal. iv, p. 145.

B. vi, D. $\frac{9}{11}$, P. 15, V. 1/5, A. $\frac{3}{11}$, C. 12, L. l. 28.

Length of head $3\frac{2}{3}$, of caudal fin 6, height of body $4\frac{1}{4}$ to $4\frac{1}{2}$ in the total length. *Eyes*—diameter 4 in the length of the head, $1\frac{3}{4}$ diameters from the end of the snout, and $\frac{3}{4}$ to 1 diameter apart. *Fins*—spines of dorsal fin not so high as the rays: caudal rounded. *Colours*—of a brownish red becoming silvery along the abdomen, a vertical blue band or spot behind the upper half of the orbit: some oblique red streaks on the head: a black spot superiorly at the base of the pectoral fin. Dorsal fin reddish with two or three rows of round yellowish spots, caudal of a similar colour but the spots irregularly disposed. Anal fin reddish.

Habitat.—Singapore and Colombo (Haly, *Taprobanian*, i, p. 165).

Add PLATYGLOSSUS ROSEUS.

Page 401. Day., Proc. Zool. Soc. 1888, p. 264.

B. vi, D. $\frac{9}{11}$, P. 14, V. 1/5, A. $\frac{3}{11}$, C. 14, L. l. 28, L. tr. $\frac{2}{10}$.

Length of head $4\frac{1}{3}$, of caudal fin $6\frac{1}{4}$, height of body $3\frac{3}{4}$ in the total length. *Eyes*—diameter $\frac{1}{4}$ length of head, $1\frac{1}{2}$ diameters from the end of snout and 1 apart. The greatest width of the head equals half its length. *Teeth*—a posterior canine. *Fins*—caudal slightly rounded: the length of the pectoral equals that of the head behind the middle of the eye: outer ventral ray somewhat elongated. *Scales*—none on the head, those on chest smaller than on the body. *Colours*—in a spirit specimen rosy, with a large black spot behind the middle of the eye and a small one between the two first dorsal spines: two narrow light bands pass from the eye to the snout: a broad orange band along the suborbital ring of bones: body with dark and narrow horizontal bands in its anterior half, while seven dark and wider bands pass from the back down the sides. A narrow light band goes from the eye to the middle of the base of the caudal fin. Basal third of caudal fin somewhat dark, with its outer edges light.

Habitat.—Kurrachee in Sind.

Page 408. Add CORIS HALEI.

Coris, sp. Haly, Taprobanian, i, 1886, p. 165.

B. vi, D. $\frac{8}{12}$, V. 1/5, A. $\frac{3}{11}$, L. l. 75, L. tr. 3/27.

"Height of body $3\frac{1}{4}$ of the total length, the length of head $\frac{1}{5}$. *Fins*—anterior dorsal spine elevated, and equal to the height of the body. *Colours*—body vinous-red, barred by eleven purplish-gray stripes: each scale with a spot of brilliant emerald green. Head orange, with violet, red-bordered stripes radiating from the eye, two of these unite to form a broad band descending from the fourth dorsal spine, past the eye and the mouth to the subopercle. A broad red band on the edge of the opercles. Dorsal fin red gray, with an orange band covered with small blue spots: anal vinous-red, with an orange border and covered with small blue spots. Caudal dark gray with large blue, black-edged spots."

Habitat.—Ceylon. A somewhat allied species seems to exist in *Coris Bleekeri*, Hubrecht, Ann. and Mag. Nat. Hist. 1876 (4) xvii, p. 214.

Page 413. Add PSEUDOSCARUS BATAVIENSIS.

Scarus bataviensis, Bleeker, Java, iv, p. 342.
Pseudoscarus bataviensis, Bleeker, Atl. Ich. i, p. 48, t. xii, f. 3; Günther, Catal. iv, p. 231.

B. v, D. $\frac{9}{10}$, P. 14, V. 1/5, A. $\frac{2}{9}$, C. 13, L. l. 25.

Length of head 4, height of body $3\frac{1}{4}$ in the total length. *Eyes*—diameter 6 in the length of the head, $2\frac{1}{2}$ diameters from the end of the snout. *Teeth*—two small ones at the corner of either jaw. *Fins*—the dorsal spines of about the same length equalling one-fourth of that of the body beneath. Caudal nearly square in the young, the outer rays produced in old examples. *Scales*—two rows on the cheeks, none covering the lower limb of the preopercle. *Colours*—head superiorly Indian red, extending to snout and throat, becoming gradually more diluted over the belly: golden green on cheeks and opercles. Eye surrounded by emerald green, passing downwards in two stripes to the upper lip and chin, a second short one behind the chin: two short branches from the hind edge of the orbit. Body olive green becoming paler below. Dorsal fin banded as follows from summit to base, cobalt, deep vinous-red, emerald green, vinous-red and cobalt. Pectoral pale orange. Ventral rosy with its outer ray blue: anal banded as follows from outer edge to base, cobalt, rose, cobalt, red and cobalt. Caudal venetian red, its upper and lower edges and three vertical bands cobalt. Every scale red at its base.

Habitat.—Arracan to the Malay Archipelago.

Page 413.   Add Pseudoscarus dussumieri.

? *Scarus dussumieri*, Cuv. and Val. xiv, p. 252; Bleeker, Batav. p. 404.
Pseudoscarus dussumieri, Bleeker, Scar. 1861, p. 13, and Atl. Ich. i, p. 46, t. viii, f. 1; Günther, Catal. iv, p. 224.

B. v, D. $\frac{9}{10}$, P. 15, V. 1/5, A. $\frac{2}{9}$, C. 13, L. l. 25.

Length of head 3½, of caudal fin 6½ in the total length. *Eyes*—diameter 6½ in the length of the head, and 3 diameters from the end of the snout. *Teeth*—small ones at the corner of either jaw. *Fins*—dorsal spines slightly increasing in length posteriorly and not so high as the rays: caudal emarginate except in the young. *Scales*—two rows on the cheeks and two scales on the preopercular limb. *Colours*—cœrulean blue, with the lower edge of the body pale rose: the centre of every scale on the blue portion being gall-stone green, as is also the upper portion of the head. Cœrulean blue bands and marks are round the orbit also radiating from it towards the forehead, the angle of the mouth, and irregularly over the cheeks. A blue band across the upper lip and another a short distance behind the lower one. Dorsal and anal fins of an orange gall-stone, having a cœrulean blue basal, and a second outer band. Caudal of a similar colour with its outer rays blue, and three broken vertical blue bands on its outer half. Pectoral rays as follows: the upper blue, the succeeding four gall-stone orange, the remainder hyaline. Ventral outer ray blue, the rest hyaline with the outer halves of the second and third rays orange gall-stone.

Habitat.—Red Sea, Persian Gulf, Arracan to the Malay Archipelago.

Page 419.   Add 2. Brotula jerdoni.

D. 126, V. 1, A. 95.

Length of head 6, height of body 5¾ in the total length. *Eyes*—in figure, diameter 3½ in the length of the head, and ⅔ of a diameter from the end of the snout. *Fins*—dorsal commences over the base of the pectoral, vertical ones confluent. *Colours*—lilac along the back becoming white beneath, a black band from the eye to the angle of the subopercle, a second from above the eye passes downwards to the base of the pectoral, which fin it crosses obliquely, a third black band commences on the occiput but soon divides into two, the upper branch going along the base of the dorsal fin, and the lower passing down a short distance and then running parallel to the first. *Fins* yellowish, three large round black spots edged with white along the upper half of the dorsal fin, which has a dark margin, as has also the anal, which, however, is externally edged with white.

Habitat.—Taken at Madras in August, and among Sir. W. Elliot's and Jerdon's illustrations is one 5·4 inches in length.

Page 419.   3. Brotula multibarbata.

? *Brotula multibarbata*, Schlegel, Fauna Japon. Poiss. p. 251, pl. cxi, f. 2; Günther, Catalog. iv, p. 371.
Geneiates ferruginosus, Tickell, MSS. with a figure.

D. C. and A. 165 (186), V. 2.

Height of body 5 (4 to 4½), length of head 5½ in the total length. Upper jaw the longer. The maxilla extends to below the hind edge of the eye, opercle ending in a spine. *Barbels*—both jaws with three on either side. *Fins*—dorsal commences above the base of the pectoral, all the vertical fins confluent. *Colours*—reddish or vinous burnt umber becoming nearly white below. Vertical fins a little darker edged with black having an outer red margin.

Habitat.—Col. Tickell procured one 4·9 inches long at Saddle Island off Kyoukphoo in Arracan in 1862, it is not so deep as shown in Schlegel's figure: Jerdon also procured one 8 inches long at Madras, the depth of which to the total length was still less, being only one-seventh: a good figure of it exists.

Page 419. Add Genus—*Fierasfer*, Cuv.

Echiodon, Thompson: *Diaphasia*, Lowe: *Oxybeles*, Richardson: *Porobronchus (young)*, Kaup.
Branchiostegals seven, pseudobranchiæ absent. Body terminating in a long and tapering tail. Gill-opening wide, the membranes united beneath the throat, but not attached to the isthmus. Gills four. The upper jaw overlapping the lower. No barbels. Cardiform teeth in the jaws, vomer, and palatines, while canines may likewise be present. Vertical fins continuous, ventrals absent. Vent under the throat. Scales, if present, minute. Air-bladder present. Pyloric appendages absent.

FIERASFER HOMEI.

Oxybeles homei, Richards. Voy. Erebus and Terror, Fishes, p. 44, pl. xliv, fig. 7-18.
    ,,   *brandesii*, Bleeker, Verh. Bat. Gen. xxiv, Chironec. p. 24 and Nat. Tyds. Ned. Ind. i, p. 276, f. 1-3.
Fierasfer homei, Kaup, Apodal Fish, p. 158; Günther, Catal. iv, p. 382.

Length of head 7¼, height of body 12 in the total length. *Eyes*—diameter ¼ of the length of the head, half a diameter from the end of the snout, and 1 diameter apart. The greatest width of the head ⅔ of its length. Snout rounded. The upper jaw the longer: the maxilla reaching to behind the posterior edge of the eye. *Teeth*—in the upper jaw in a rather widely-set, recurved row, and a canine-form one near the centre of the jaws: an outer row of small teeth. In two rows in the lower jaw the outer being the larger and somewhat curved, they are largest near the symphysis: 2 or 3 rows on the palatines: 2 large ones, placed one before the other on the vomer, and surrounded by smaller teeth. Vent in front of a line from the base of the pectoral fin. *Scales*—absent. *Fins*—vertical ones enveloped in skin: the dorsal commencing the length of the head behind the front edge of the eyes: the anal beginning behind the vent. Pectoral as long as the head behind the eyes. *Colours*—yellowish-red, a silvery band going from the upper edge of the opercles along the first fourth of the body: opercles silvery: end of tail with some black reticulations.

Habitat.—An example 4·8 inches long received from Madras. It is found in the Malay Archipelago and the Australian Seas.

Page 450. Add LEIOCASSIS FLUVIATILIS.

Duxordia fluviatilis, Tickell, MSS. p. 338, c. fig.

    B. vi, D-½/0, P. 1/7, V. 6, A. 11, C. 18.

Length of head 4¼, of caudal fin 5¼, height of body 5 in the total length. *Eyes*—rather small, high up and in the anterior half of the head. The greatest width of the head equals two-thirds of its length. There is a moderate rise from the snout to the base of the dorsal fin. Upper jaw the longer, upper surface of head smooth. *Barbels*—a maxillary pair reaching to the posterior edge of the orbit, no others were detected. *Teeth*—in an uninterrupted villiform mass across the palate. *Fins*—dorsal spine smooth and nearly as long as the fourth ray which is 4/5 as high as the body below it. Adipose dorsal rather long, commencing a short distance behind the base of the rayed fin. Pectoral spine denticulated internally. Caudal forked. *Colours*—yellowish horny with darker shades of olive brown on the snout and along the back, also some cloudy markings. A large black blotch on the lateral-line above the anal fin, another between the pectoral and first dorsal. Tip of dorsal and ends of both caudal lobes black.

Habitat.—Col. Tickell obtained four examples, the largest 3½ inches long from the Anin, a stream rising near Weywoon, Wagroo in the Tenasserim Provinces.

Page 474. Add Genus—*Akysis*, Bleeker.

Body somewhat elongated: head broad, and covered with soft skin. Gill-openings of moderate width, the membranes stretching across the isthmus, and being slightly notched posteriorly. Mouth terminal: the upper jaw slightly the longer. Nostrils, the anterior one with slightly tubular edges, posterior with a barbel before it. Barbels eight. Eyes small. Villiform teeth in the jaws, none on the palate. A short dorsal fin with one spine and five rays: pectorals horizontal: ventral with six rays: caudal emarginate or forked. Lateral-line present. Skin tubercular.

Geographical distribution.—From the Tenasserim Provinces to the Malay Archipelago.

AKYSIS PICTUS.

Günther, Ann. and Mag. N. H. (5) xi, p. 1883, p. 138.

D. ½/0, P. 1/7, V. 6, A. 9.

Head broader than deep. *Eyes*—wide apart, and twice as distant from the gill-opening as from the end of the snout. The distance of the anterior nostrils apart equals about half the length of the snout, while the interspace between the anterior and posterior nostrils equals half that present between the front pair. *Barbels*—nasal half as long as the head, the maxillary reaching to the origin of the dorsal fin, the outer mandibular ones to the axil of the pectoral, while the inner ones are shorter. *Fins*—dorsal commences midway between the snout and the adipose fin, its spine comparatively strong. Anal arises nearer the root of the caudal than that of the pectoral. Caudal emarginate: pectoral extending a little beyond the origin of the dorsal, its spine strong and entire: ventrals reaching the vent. *Colours* — head grayish with minute black spots, body anteriorly black which is contracted into an irregular band that runs along the middle of the posterior part of the body and tail. Dorsal fin with a black band covering all but its front corner and upper edge: caudal and pectoral banded.

Habitat.—Tenasserim to 45 millim long.

Page 475. OLYRA LONGICAUDA. Add synonym.
Olyra elongata, Günther, Annals and Mag. Nat. Hist. p.

Page 503. Add Family—GALAXIDÆ.

Body more or less elongated: abdomen rounded. Pseudobranchiæ absent. Edge of upper jaw mainly formed by the premaxillaries. Dorsal fin opposite to the anal, no adipose fin. Air-bladder large and simple. Pyloric appendages few. The ova pass into the abdominal cavity before exclusion.

Genus 1.—GALAXIAS, *Cuvier*.

Mesites, Jenyns.

Definition as in family. Conical teeth in both jaws, vomer and palatine bones, and large ones on the tongue.

Habitat.—Southern portion of South America, Australia, New Zealand, and observed to live in fresh waters : this Indian form was from the littoral district.

GALAXIAS INDICUS.

B. ix, D. 13, P. 10, V. 8, A. 18, C. 15.

Length of head 8¼, of caudal fin 8¼, height of body 11 in the total length. *Eyes*—3½ diameters in the length of the head and 1¾ from the end of the snout. Body elongated and flattened, with a rounded abdomen. *Teeth*—five conical ones in the lower jaw, vomer and palatine bones, and some larger ones on the tongue. *Fins*—ventral well developed and arising midway between the hind edge of the eye and the posterior extremity of the base of the anal fin. Dorsal fin commences opposite the origin of the anal, and in about the commencement of the last third of the total length it is highest in front, and the extent of free portion of the tail behind it equals about 1½ in the length of its base. Caudal forked.

Habitat.—Littoral districts of Bengal and Madras, attaining about 2 inches in length.

Among the drawings of the late Sir Walter Elliot is one of a small fish, a little over 1 inch in length, and a magnified copy nearly four times that size. It was taken at Waltair, April 8th, 1853. Its form is deeper than the foregoing, while it has D. 17, A. 24. No ventral fins are shown, and the vent is placed in the centre of the length of the body. Dorsal fin commences slightly in advance of the anal and in the commencement of last third of the total length : caudal forked. *Colours*—a row of black spots along the edge of the abdomen. Sufficient details are not given to render one able to decide on the position it should hold. In the absence of ventral fins, which may have been overlooked, it somewhat approaches the *Leucopsarion Petersii* of Hilgondorf.

Page 520. Add EXOCŒTUS ALTIPINNIS.

Cav. and Val. xix, p. 109, pl. 560; Bleeker, Atl. Ich. vi, Scomb. t. i, f. 3 (*ventrals too short*); Day, Proc. Zool. Soc. 1888, page

Exocœtus katopron, Bleeker, Atl. Ich. vi, p. 72.

B. xi, D. 13, P. 14-15, V. 6, A. 10, C. 14, L. l. 52; L. tr. 7-8/2.

Length of head $5\frac{1}{4}$ to $5\frac{1}{2}$, of caudal fin $4\frac{3}{4}$ to 5, height of body 7 to $7\frac{1}{2}$ in the total length. *Eyes*—diameter $2\frac{3}{4}$ in the length of the head, $\frac{3}{4}$ of a diameter from the end of the snout, and rather more than 1 apart. Interorbital space flat or rather concave. *Barbels*—absent. *Teeth*—rudimentary. *Fins*—dorsal commences between the hind edge of the orbit and the end of the lower caudal lobe, anteriorly it is two-thirds as high as the body. Ventrals commence midway between the hind edge of the eye and the base of the caudal fin, reaching to the end of the base of the anal. Anal begins on a line below the middle of the dorsal fin. *Scales*—28 rows between the occiput and the base of the dorsal fin. *Colours*—bluish, becoming silvery along the abdomen: pectoral nearly black with the first ray white, and an oblique wide white band crossing from its outer edge to a little in front of its base. In one specimen the ventral is black tipped: caudal grayish.

Habitat.—Two specimens up to $11\frac{1}{2}$ inches long received from Bombay: it extends to the Malay Archipelago.

Page 549. CIRRHINA FULUNGEE. Add synonym.

Gobio angrioides, Jerdon.

Page 551. SCAPHIODON IRREGULARIS. Add synonym.

Cirrhina afghana, Günther, Trans. Linn. Soc. 1887.

Not only does this fish differ from those of the genus *Cirrhina* in the character of its mouth, but it likewise possesses a serrated osseous ray in the dorsal fin.

Page 564. BARBUS TOR.

This species is found in Ceylon, according to Haly.

Page 582. Add Genus—*Acanthonotus*, Tickell (MSS.).

Mouth arched, anterior: barbels absent, eyes without adipose lids. Dorsal fin rather short, commencing slightly anterior to the root of the ventral, its osseous ray being strong, serrated and preceded at its base by a small forwardly-directed spine: anal short. Scales large, no enlarged row at base of anal fin. Lateral-line complete and continued to opposite the centre of the base of the caudal.

1. ACANTHONOTUS ARGENTEUS.

Tickell, MSS. page 49, with a figure.

D. 9 ($\frac{1}{8}$), P. 14, V. 8, A. 7, C. 18, L. l. 30.

Length of head as delineated 6, of caudal fin $3\frac{2}{3}$, height of body $3\frac{1}{2}$ in the total length. *Eyes*—diameter $3\frac{1}{2}$ in the length of the head, 1 diameter from the end of the snout. Snout blunt, rather overhanging the mouth, body compressed: profile with a considerable rise from snout to base of dorsal fin. *Fins*—dorsal spine strong and posteriorly serrated, caudal deeply forked, its lobes acutely pointed. A small horizontal spine in front of the dorsal fin pointing forwards and scarcely protruding from beneath the skin. *Lateral-line*—complete. *Colours*—brilliant silvery with lilac and blue shades and a tinge of olive-yellow on the back. Dorsal fin orange-scarlet superiorly bordered with black except on the last two rays, the other fins lemon-yellow. Dorsal ridge black in its upper portion.

Habitat.—Very common in the streams of the interior of the Tenasserim district, the largest obtained being about 5·4 inches in length.

Page 587. Add ROHTEE CUNMA.

Abramis cunma, Tickell, MSS. p. 53, c. fig.

B. iii, D. 12 ($\frac{3}{9}$), P. 13, V. 10, A. 30, C. 18, L. l. 44.

Length of head $5\frac{1}{2}$, of caudal fin $4\frac{1}{4}$, height of body 3 in the total length. *Eyes*—diameter 3 in the length of the head, 1 diameter from the end of the snout and also apart. Profile over nape slightly concave, a considerable rise from snout to base of dorsal fin: snout somewhat obtuse, upper jaw the longer. *Fins*—dorsal anteriorly two-thirds as high as the body below it, commencing somewhat nearer the snout than the base of the caudal fin, its spine weak and longer than the head, not serrated. Caudal deeply forked, lower lobe somewhat the longer. *Lateral-line*—strongly marked in its first four scales. *Colours*—olive, superiorly becoming silvery on the sides and below : a brassy tinge along the lateral-line and over the cheeks and gill-covers. Fins amber : dorsal and caudal with a narrow black edge.

Habitat.—Colonel Tickell procured it at Moulmein, where he found it to be common. His figure is 5¼ inches long.

Page 652. Family CHIROCENTRIDÆ. After "intestinal canal short," add "and furnished with spiral folds."

Page 679. Add SYNGNATHUS CONSPICILLATUS.

Syngnathus fasciatus, Gray, Ind. Zool. c. fig. (not Risso).
    "   *conspicillatus*, Jenyns, Voy. Beagle, Fish. p. 147, pl. xxvii, f. 4; Günther, Catal. viii, p. 174.
Syngnathus hæmatopterus, Bleeker, Nat. Tyds. Ned. Ind. ii, p. 258.
Corythoichthys fasciatus, Kaup, Lophob. p. 25; Kner, Novara Fisch. p. 391.

D. 29-32, P. 14, A. 3, C. 10, osseous rings 16-17 + 34-37.

Length of head about 10½ in the total length : tail portion more than twice as long as that of the body. Snout slender. The upper profile of the head rises abruptly above the eyes. Opercle crossed by a ridge : a median ridge on the occiput and nuchal shields : a supraorbital ridge which is continued along each side of the crown. Body slightly deeper than broad, ridges well developed : egg pouch not quite half so long as the tail. *Fins*— the dorsal commences on the anal or first caudal ring. *Colours*—trunk grayish-brown, with deep brown interrupted transverse bands, and sometimes large white intermediate spots. Fine brown lines on the head, and a band below the eye over the lower side of the opercle. Dorsal fin a little spotted.

Habitat.—From the east coast of Africa to the Pacific Ocean. The Colombo Museum possesses five specimens procured from Jaffna in the northern portion of Ceylon (Haly, *Taprobanian* i, 1886, p. 165).

Page 692. Add BALISTES BUNIVA.

Balistes niger, Osbeck, Voy. China, ii, p. 93 (not Linn.); Bl. taf. 152, f. 2; Bl. Schn. p. 472; Lacép. i, p. 370, pl. xviii, f. 1; Richards. Voy. Samarang Fishes, p. 21, pl. vi, fig. 1-4, and Ich. China, p. 201; Günther, Fish Zanzibar, p. 135, pl. xix, f. 1.
Balistes buniva, Lacép. v, p. 669, pl. xxi, f. 1; Günther, Catal. viii, p. 227.
    "   *pictus*, Poey, Proc. Ac. Nat. Sc. Phil. 1863, p. 180.
Melichthys ringens, Bleeker, Atl. Ich. v, p. 108, Balistes, pl. cxxx, f. 2.

B. vi, D. 2-3/31-33, P. 15, A. 28-30, L. 1, 53 (55 Bleeker).

Length of head 4, of caudal fin 8, height of body nearly half of the total length. *Eyes*—5 diameters in the length of the head, and 3½ from the end of the snout: a groove before the eye. *Teeth*—white, even, and incisor-like. *Fins*—first spine of dorsal fin very strong : caudal posteriorly with an S-shaped outline, in old specimens the lobes are slightly produced. *Scales*—enlarged osseous plates behind the gill-opening. Seven or eight raised and spiny lines on the side of the tail. *Colours*—black with a raised white line along the bases of both the second dorsal and anal fins.

Habitat.—Tropical parts of Atlantic, Indian, and Pacific Oceans. It has been captured in Zanzibar (Playfair) and Ceylon (Haly), where it appears to be common.

Page 693. Add MONACANTHUS TORMENTOSUS.

Balistes tormentosus, Linn. Syst. Nat. i, p. 405; Gronov. ed. Gray, p. 34; La Cépède, i, pp. 333, 359.
Monacanthus tormentosus, Cuv. Regne Anim : Bleeker, Atl. Ich. v, p. 127, Balistes pl. vi, fig. i, *male*, and pl. xvi, fig. i, *female :* Günther, Catal. viii, p. 238.
    "   *hajam*, Bleeker, l. c. p. 126, Balistes, pl. i, fig. 1, *female*, and pl. xvi, fig. 1, 3, *male*.
    "   *trichurus*, Bleeker, Nat. Tyds. Ned. Ind. iv, p. 125.
    "   *helleri*, Steind. Sitz. Ab. Wiss Wien. 1867, iv, p. 712, t. iii, f. 3.

B. vi, D.-1/26-29, P. 11, A. 25-27, C. 10.

Length of head 4, of caudal fin 5⅓, height of body 2⅓ in the total length. *Eyes*— situated high up in the posterior half of the head over the gill-opening and 2 diameters from the end of the snout. Body somewhat elevated, profile from snout to first dorsal fin somewhat concave. *Fins*—Dorsal spine situated over the hind quarter of the eye, strong and about as long as the head, armed posteriorly with a double row of recurved spines. Ventral spine movable with small curved spines posteriorly, second dorsal and anal fins low : caudal rounded. *Scales*—small, each with 4 or 5 marginal spines, while the male on either side of the tail has a long patch of short setiform spines sometimes absent from the female. Some short fleshy tentacles on the side of the body. *Colours*—brown, spotted and marbled with black, a light band along the anterior half of the body. Caudal fin with two dark vertical bands.

Habitat.—Obtained by Mr. Haly in August, 1888, at Karativoe, Ceylon ; is found in the Malay Archipelago, Chinese and Australian seas.

Page 708. The Colombo Museum sustained a great loss during my absence last year : a small sun-fish, *Orthagoriscus*, was brought for sale, but was unfortunately rejected (Haly, in the *Taprobanian*, vol. ii, 1888, p. 165). This may have been the widely ranging *O. mola* not uncommon off the British coast, and which has been taken in New South Wales, &c., and of which Klunzinger seems to have obtained a specimen in the Red Sea : or it may be the form found at Amboina which was described and figured by Bleeker in 1873 as *O. oxyuropterus* : or possibly a nondescript.

Page 709. Add DIODON MACULATUS.

Diodon tacheté, Lacép. ii, p. 13.
    ,, *novem-maculatus*, Cuvier, c. fig. ; Bleeker, Nat. Tyds. Ned. Ind. iii, p. 567.
    ,, *sex-maculatus* and *quadrimaculatus*, Cuv. c. fig. ; Kaup, pp. 229, 227.
    ,, *spinosissimus*, Kaup, p. 228 (not Cuv.).
Paradiodon novem-maculatus, Bleeker, Atl. Ich. v, p. 57, Gym. pl. ii, f. 3.
    ,, *quadri-maculatus*, Bleeker, l. c. p. 58, pl. viii, f. 2.
Diodon maculatus, Günther, Cat. viii, p. 307.

B. vi, D. 2/13, P. 23, A. 2/12, C. 7.

Length of head 2¾ to 3¼ in the length of the body. *Eyes*—diameter 3½ to 4 in the length of the head. Sometimes tentacles above the orbit and on the lower side of the head, and on the back. Spines of varying lengths, from 16 to 19 between the snout and the dorsal fin : there are generally only two or three posterior to the dorsal fin. The roots of these spines are long and strong, and have a distinct ridge along their basal portions. *Colours*—large black yellow-edged blotches on the body of various shapes, and often small black spots. The large black spots are in some cases badly defined.

Habitat.—Tropical portions of the Atlantic, Indian Ocean, and Archipelago, also the Pacific. Found in the Gulf of Manaar by Sir Walter Elliot, and in Ceylon by Haly.

Page 716. Add CARCHARIAS MURRAYI.

Günther, Ann. Mag. Nat. Hist. (5), xi, p. 137.

Snout short and obtuse : the distance between the mouth and the end of the snout being less than that between the inner angles of the nostrils. Nostrils nearly midway between the end of the snout and the mouth. *Teeth*—in the upper jaw of moderate size, the anterior equilateral, rather longer than broad, those on the side oblique, with their posterior edges concave, and both sides finely serrated : twenty-nine rows in the lower jaw, lanceolate, their edges smooth, with a broad base, two-rooted, and some with an additional minute lobe. *Fins*—first dorsal commences opposite the axil of the pectoral : the second only one-third of the size of the first, but larger than the anal, which is small : origin of anal behind that of the second dorsal. Pectoral large, exceeding the distance between the first gill-opening and the end of the snout, the length of its hind margin only one-fourth of that of its outer. Caudal of moderate size, rather more than the distance between the two dorsal fins. *Colours*—uniform, top of first dorsal may have been black.

Habitat.—Kurrachee, where an example 6 feet 8 inches long was captured. The specimen is stuffed, and not in a good condition. It is very closely allied to *C. elioti*, from which it differs in the smaller size of the second dorsal and anal fins, but is probably only a variety.

Page 720. Add ZYGÆNA MOKARRAN.

Rüppell, N. W. Fisch. 1835, p. 66, t. xvii, t. 3 ; Günther, Catal. viii, p. 383 ; Day, Ann. and Mag. N. H. (5) xx, 1887, p. 389.

Sphyrna mokarran, Müll. and Henle, Plagios. p. 54.

Zygœna dissimilis, Murray, Annal. and Mag. N. H. (5) xx, 1887, p. 304.

Anterior edge of head nearly straight, and forming a more or less right angle with its lateral margin. Length of the hind edge of one of the lobes equal to or rather exceeding its width near the eye. *Eyes*—nostrils near them: but no groove running along the front edge of the head. *Teeth*—oblique, as broad at their base as long, with an indistinct lateral notch, and serrated on both edges. *Colours*—brownish gray, becoming white beneath.

Habitat.—Red Sea to Kurrachee where one, a little over 10 feet in length, was captured in April, 1884.

Page 722.   Add LAMNA GÜNTHERI, *Murray.*

Murray, Ann. and Mag. Nat. Hist. (5), xiii, p. 349.

Said to differ from *L. spallansanii* in having $\frac{3}{2}$ teeth on either side, and the dorsal fin being a little further behind the base of the pectoral.

By an error in transcribing (Ann. and Mag. Nat. Hist. 1887) I placed this shark instead of *Carcharias murrayi*, Günther, as a synonym to *Carcharias ellioti*. It occurred owing to having noted, after having examined the type that Murray's shark from Kurrachee, that it seemed to be identical with Elliot's shark from the coasts of India.

Habitat.—Kurrachee.

Page 722.   Add Genus 2—ODONTASPIS, *Agassiz.*

Triglochis, Müller and Henle.

Spiracles minute and above the angle of the mouth. No nictitating membrane. Mouth wide and crescent-shaped. Teeth large, awl-shaped, and with one or two cusps at the base. Gill-openings of moderate size. Two spineless dorsal fins, the first opposite the interspace between the pectoral and ventral: the second dorsal and anal not much smaller than the first dorsal. A pit present or absent at the root of the caudal fin.

Geographical distribution.—Temperate and tropical seas.

1. ODONTASPIS TRICUSPIDATUS.

Carcharias tricuspidatus, Day, Fish. India, p. 713, pl. clxxxvi, fig. 1.

Dundanee, Sind.

This fish was formerly placed as a *Carcharias*, owing to the presence of a pit at the root of the caudal fin as observed at page 722.

Genus 8.—ALOPIAS, *Rafinesque.*

Mouth crescentic. No membrana nictitans to the eye. Spiracles minute, close behind the orbit. Teeth of rather small size, flattened and triangular, having smooth edges. Gill-openings of medium size. The first dorsal fin spineless, inserted above the interspace between the pectoral and ventral fins: the second dorsal above the interspace between the ventral and anal, the latter being small. Caudal very long, with a pit at its commencement. No keel on the side of the tail.

1. ALOPIAS VULPES.

Squalus vulpes, Gmel. Linn. p. 1496; Lacépède, i, p. 267; Bl. Schn. p. 127.

Carcharias vulpes, Cuv. Règne Anim.

Alopias vulpes, Bonap. Fauna Ital. Pesc. iii, p. 66, c. fig.; Müller and Henle, p. 74, pl. xxxv, f. 1 (*teeth*); Gray, Catal. Chond. p. 64; Day, Fish. Great Britain and Ireland, ii, p. 300 (see synon.).

Alopecias vulpes, Yarrell, Brit. Fish. (ed. 3), ii, p. 512, c. fig.; Günther, Catal. viii, p. 393.

Body fusiform, gradually decreasing in size to the caudal fin, the great length of which is about half of the total. *Eyes*—rather large. Nostrils beneath and nearer the anterior border of the mouth than the end of the snout. Gill-opening of median size, the last two being over the pectoral fin. *Teeth*—about $\frac{22}{19} + \frac{22}{19}$, the third or fourth tooth on either side of the centre of the upper jaw smaller than the others.

Habitat.—Atlantic Ocean on both shores. One from the Cape of Good Hope is in the Paris Museum, and Mr. Haly in the *Tabrobanian*, 1886, i, p. 167, records one 8 ft. 8 in. in length from Ceylon, having been procured from the Colombo market, February, 1884, where it was quite unknown to the fishermen. It is also found in the Mediterranean, and has been obtained from San Francisco Bay, California, and New Zealand.

Page 723. Add

FAMILY—RHINODONTIDÆ.

Spiricales minute : no nictitating membrane. Gill-openings wide. Two spineless dorsal fins, the origin of the first somewhat in advance of the ventrals: the second small, placed nearly opposite the anal : lower caudal lobe well developed. A keel along the side of the tail. A pit at the root of the caudal fin.

Genus I.—RHINODON, *Smith.*

Definition as in the family. Mouth and nostrils near the extremity of the snout. Teeth small and conical. Gill-rakers similar to those of the basking-shark of Northern seas.

Geographical distribution.—Ceylon and Seychelles to the Cape of Good Hope. Specimens are said to have exceeded fifty and even seventy feet in length. It is a harmless form.

RHINODON TYPICUS.

Smith, Illus. S. African Fish, pl. 26; Müller and Henle, p. 77, t. xxxv, f. 2 (teeth); Dumeril, Elasm. p. 428; Haly, Ann. and Mag. N. H. (5), xii, p. 48.
Snout broad, flat, and short. *Eyes*—small. Upper jaw with a long labial fold. *Colours*—brownish white dots and narrow transverse lines.
Habitat.—One example recorded from the west coast of Ceylon.

Page 725. Add 2. GINGLYMOSTOMA CONCOLOR.

Nebrius concolor, Rüpp. N. W. Fische, p. 62, t. xvii, f. 2.
Ginglymostoma concolor, Cantor, Mal. Fish. p. 395; Günther, Catal. viii, p. 409; Klunz. Synopsis F. R. M. 1871, p. 672.
Ginglymostoma rüppellii, Bleeker, Verh. Bat. Gen. xxiv, Plagios. p. 91; Dumeril, Elasm. p. 334.

Snout short. The nasal cirrus nearly reaches the lower lip. *Teeth*—in three rows, with one central and four or five lateral cusps, having serrated edges. *Fins*—dorsal, pectoral, and anal fins with pointed angles. Second dorsal much smaller than the first, and placed nearly opposite to but larger than the anal. Caudal fin one-third of the total length.

Habitat.—Red Sea, through those of India to the Malay Archipelago.

Page 729. Add 4. PRISTIS PECTINATUS.

Latham, Trans. Linn. Soc. 1794, ii, p. 278, pl. xxvi, f. 2 (snout); Bl. Schn. p. 351, pl. lxx, f. 1; Müll. and Henle, p. 109; Blyth, Journ. As. Soc. Beng. 1860, p. 36; Duméril, Elasmobranchs, p. 475; Günther, Catal. viii, p. 437; Klunz. Synop. F. R. M. 1871, p. 673.

Squalus scio, Lacép. i, p. 286, pl. viii.

Rostrum nearly twice as wide at its base as at its termination, armed with from 24 to 27 pairs of teeth which are generally long and somewhat strong and not placed opposite one another, while they may be directed somewhat posteriorly. Anteriorly the interspace between each tooth equals about the width of their base, but among the most posterior ones it becomes double that distance. *Fins*—first dorsal commences opposite the ventral, the second dorsal about of equal size to the first. No lower caudal lobe. *Colours*—sandy-brown becoming lighter beneath.
Habitat.—Red Sea, through the Indian Ocean.

Page 732. Add RHINOBATUS COLUMNÆ.

Rhinobatus (Syrrhina) columnæ, Müller and Henle, p. 113 : Duméril, Elasm. p. 486.
   ,,   ,,   ,,   *annulatus*, Müll. and Hen. p. 116 : Smith, Illus. Zool. S. Afri. Pisces, pl. xvi : Duméril, l. c. p. 487.

Raja rhinobatus, Gronov. ed. Gray, p. 10.
Rhinobatus (Syrrhina) polyophthalmus, Bleeker, Japan, p. 129.

Snout rather elongated : the distance between the outer angles of the nostrils equals two-fifths of the extent preoral portion of the snout. Anterior nasal valve is connected to a fold of skin passing towards the median line and so nearly joins that of the opposite side. The upper vortical ridges are convergent in front. Back finely granular with a medium row of small tubercles. *Colours*—brown, young examples have a white snout.

Habitat.—Mediterranean and the Indian and Atlantic Oceans.

Page 745. Erase Genus CERATOPTERA.

Ceratoptera ehrenbergii.

The figure must, I think, refer to an abnormal condition of *Astrape dipterygia,* as I find such a form of monstrosity more common among European rays and skates than I had formerly reason for supposing.

Page 729. Add

SUB-CLASS—LEPTOCARDII.

Skeleton semicartilaginous and notochordal : destitute of jaws or ribs. Brain absent. Blood colourless and distributed by pulsating sinuses. Respiratory and abdominal cavities confluent : numerous branchial clefts and the water discharged by an opening in front of the vent.

FAMILY I.—CIRROSTOMI.

An elongated compressed body, having a low and rayless dorsal fin, continued round the tail past the vent to the respiratory opening. Mouth a longitudinal slit on the inferior surface, and with cirri. Eyes rudimentary. Vent near the end of the tail.

Genus 1—BRANCHIOSTOMA, *Costa.*

Amphioxus, Yarrell.
Definition as in the family.

One or more species of this genus are common around the waters of India, Burma, Ceylon, and the Andaman Islands.

INDEX TO SUPPLEMENT, 1888.

Abramis canma, 807
Acanthoclinus, 798
Acanthoclinus indicus, 798
Acanthonotus, 807
Acanthonotus argentens, 807
Acanthurus annularis, 789
Acanthurus argenteus, 789
Acanthurus Blochii, 789
Acanthurus galm, 789
Acanthurus mata, 789
Acanthurus matoides, 789
Acanthurus melanurus, 789
Acanthurus nigricans, 789
Acanthurus striatus, 789
Acanthurus strigosus, 789
Acanthurus tennentii, 788
Acanthurus tristis, 788
Acanthurus xanthopterus, 789
Acronurus lineolatus, 789
acutipinnis, Gobius, 793
acutirostris, Serranus, 780
acutus, Lethrinus, 787
adusta, Myripristis, 788
adusta, Pseudochromis, 791
afghana, Cirrhina, 807
Akysis, 805
Akysis pictus, 806
albofasciatus, Pomacentrus, 801
alboguttatus, Salarias, 798
albovinctus, Glyphidodon, 801
Alopecias vulpes, 810
Alopias, 810
Alopias vulpes, 810
altipinnis, Exocœtus, 807
altispinis, Gerres, 786
altivelis, Cromileptes, 779
altivelis, Serranus, 780
Ambassis baculis, 784
Ambassis myops, 784
Ambassis notatus, 784
Ambassis ranga, 784
Amphioxus, 812
Amphiprion bifasciata, 800
Amphiprion intermedius, 800
Amphiprion seba, 800
Amphiprion trifasciatum, 800
angrioides, Gobio, 807
angularis, Serranus, 780
annularis, Acanthurus, 789
annulatus, Rhinobatus, 811
Anthias argus, 780
Anthias hæmruhr, 781
Anthias multidens, 782
Antika doondinwah, 789
antjerius, Glyphidodon, 801
antjerius, Glyphidodontops, 801

Apharaeus, 782
Apharaeus cærulescens, 782
Apharaeus furcatus, 782
Apharaeus rutilans, 782
Apogon amboro, 784
Apogon bifasciatus, 784
Apogon ellioti, 784
Apogon endekatænia, 784
Apogon fasciatus, 784
Apogon lineolatus, 785
Apogon macropterus, 785
Apogon maximus, 784
Apogon pœcilopterus, 785
Apogon thurstoni, 784
Apogon tickelli, 785
aporos, Ophiocara, 795
Aprion pristopoma, 782
arabica, Perca, 785
arabica, Trigla, 791
arabicus, Cheilodipterus, 785
armatura, Apogon, 784
arenatus, Salarias, 798
areolata, Perca, 780
areolatus, Serranus, 780
argentaria, Gazza, 790
argentarius, Zeus, 790
argenteus, Acanthurus, 789
argenteus, Acanthonotus, 807
argentimaculatus, Lutjanus, 783
argus, Anthias, 780
aries, Chrysophrys, 788
ater, Glyphidodon, 801
Astrape diptorygia, 812
aurigan, Trichiurus, 788
aurolineatus, Mesoprion, 788

baculis, Ambassis, 784
Balistes buniva, 808
Balistes niger, 808
Balistes pictus, 808
Balistes tormentosus, 808
bataviensis, Pseudoscarus, 803
bataviensis, Scarus, 803
Bengalensis, Holocentrus, 788
Bengalensis, Lutjanus, 783
berda, Chrysophrys, 788
bicolor, Salarias, 798
bifasciata, Amphiprion, 800
bifasciatus, Apogon, 784
bifasciatus, Prochilus, 800
bilineatus, Pomacentrus, 801
bilunulatus, Cossyphus, 802
bilunulatus, Labrus, 802
bipinnulatus, Seriolichthys, 789
bixanthopterus, Caranx, 789
Blennius leopardus, 796

Blennius steindachneri, 796
blochii, Acanthurus, 789
blochii, Dentex, 786
blochii, Priacanthus, 783
Bodian cuvieri, 786
botcho, Myripristis, 788
Branchiostoma, 812
brandesii, Oxybeles, 805
brevis, Salarias, 796
Brotula jerdoni, 804
Brotula multibarbata, 804
buniva, Balistes, 808

cæruleolineata, Mesoprion, 783
cærulescens, Apharaeus, 782
calcarifer, Lates, 779
canescens, Chaetodon, 786
canescens, Zanclus, 786
caninus, Caranx, 789
Carangoides hemigymnostethus, 789
Carangoides telamparoides, 789
Carangus marginatus, 789
Caranx bixanthopterus, 789
Caranx caninus, 789
Caranx edentulus, 789
Caranx ferdau, 789
Caranx flavo-cæruleus, 789
Caranx hippos, 789
Caranx impudicus, 789
Caranx jayakari, 789
Caranx kurra, 789
Caranx malabaricus, 789
Caranx melampygus, 789
Caranx nigrescens, 789
Caranx parapistes, 789
Caranx rüppellii, 789
Caranx speciosus, 789
Caranx stellatus, 789
Caranx venator, 789
Carcharias ellioti, 809, 810
Carcharias murrayi, 809, 810
Carcharias tricuspidatus, 810
Carcharias vulpes, 810
carinatus, Mugil, 800
caudolineatus, Salarias, 798
cavifrons, Pseudolates, 779
centrognathum, Gnathocon-
    trum, 786
centrognathus, Zanclus, 786
Centropristis pristopoma, 782
Cepola indica, 796
Ceratoptera, 812
Ceratoptera ehrenbergii, 812
chabrolii, Cossyphus, 802
Chaetodon canescens, 786

Chaetodon citrinellus, 786
Chaetodon fasciatus, 786
Chaetodon flavus, 786
Chaetodon guttatissimus, 786
Chaetodon lividus, 801
Chaetodon lunula, 786
Chaetodon miliaris, 786
Chaetodon nudus, 786
Chaetodon ocellatus, 786
Chaetodon oligacanthus, 786
Chaetodon quadrifasciatus, 786
Chaetodon tau nigrum, 786
Chaetodon trifasciatus, 786
Chaetodon vittatus, 786
Chaetodon wiebeli, 786
Chaetopterus pristipoma, 782
Cheilodipterus arabicus, 785
Cheilodipterus lineatus, 785
Cheilodipterus macrodon, 785
Cheilodipterus panijius, 791
Cheilinus undulatus, 802
Cheilinus undulatus, 802
Chilinoprista, Scorpæna, 788
Chirocentridæ, 788
chlorostigma, Gobius, 793
Chorinemus lysan, 789
Chorinemus mauritiana, 789
Chorinemus moadetta, 789
Chorinemus orientalis, 789
Chorinemus sancti-petri, 789
Chorinemus tol, 789
Chorinemus toloo, 789
Chrysophrys aries, 788
Chrysophrys berda, 788
Chrysophrys cuvieri, 788
Chrysophrys datnia, 788
Chrysophrys grandoculis, 787
Chrysophrys haffara, 788
chrysurus, Pomacentrus, 801
cincta, Pterois, 788
cingulus, Glyphidodon, 801
Cirrhina afghana, 807
Cirrhina fulunga, 807
Cirrhites fasciatus, 788
Cirrhitichthys fasciatus, 788
cirrhostomus, Mugil, 800
Citharoodus vittatus, 786
citrinellus, Chaetodon, 786
coccinicauda, Malacocanthus,
    791
columnæ, Rhinobatos, 811
compressus, Grammistes, 783
concolor, Ginglymostoma, 811
concolor, Nebrius, 811
conspicillatus, Syngnathus, 788
coppingeri, Trachynotus, 790
Coracinus vittatus, 800

Coris hulei, 808
cornutus, Zanclus, 786
CorysLion orientalis, 792
Corythoichthys fasciatus, 808
Cossyphus bilunulatus, 802
Cossyphus chabrolii, 802
Cossyphus maldat, 802
Cossyphus micrurus, 802
Crassilabrus undulatus, 802
crenilabris, Mugil, 800
cristatus, Trichiurus, 788
Cristiceps, 799
Cristiceps halei, 799
Cromileptes altivelis, 770
cruentipinnis, Salarias, 797
Cul nachooli, 794
cumma, Abramis, 807
cumma, Rohtee, 807
cuvieri, Bodian, 785
cuvieri, Chrysophrys, 788
cuvieri, Diagramma, 785
cuvieri, Plectorhynchus, 785
cuvieri, Sparus, 788
cyaneus, Glyphidodon, 801
cylindrica, Percis, 790

Dactylopterus orientalis, 792
Dascyllus niger, 801
Dascyllus trimaculatus, 801
Dascyllus unicolor, 801
Datnia, 785
datnia, Chrysophrys, 788
Datnioides polota, 786
Datnioides quadrifasciatus, 786
Decapterus Russellii, 789
Dentex blochii, 786
dentex, Equula, 790
Dentex filamentosus, 786
dentex, Lutjanus, 782
dentex, Mesoprion, 782
Dentex pristopoma, 782
Dentex tæniopterus, 786
diacanthus, Serranus, 780
Diacope macolor, 788
Diacope nigra, 783
Diagramma cuvieri, 785
Diagramma griseum, 785
Diagramma jayakari, 785
Diagramma lessonii, 785
Diagramma sobæ, 785
Diaphasia, 805
Diodon maculatus, 809
Diodon novem-maculatus, 809
Diodon quadrimaculatus, 809
Diodon sex-maculatus, 809
Diodon spinosissimus, 809
Diodon tacheté, 809
dipterygia, Astrape, 812
dispar, Glyphidodon, 801
dissimilis, Zygæna, 810
Domina, Sillago, 791
Dundanec, 810
dussumieri, Pseudoscarus, 804
dussumieri, Scarus, 804
Duxordia fluviatilis, 805

Echiodon, 805
odontulus, Caranx, 789
ahrenbergi, Ceratoptera, 812
ahrenbergii, Mesoprion, 783
Elagatis pinnulatus, 789
Eleotris Elliotti, 794
Eleotris hoedtii, 794
Eleotris lineato-oculatus, 794
Eleotris macrocephalus, 795
Eleotris macrolepidota, 794
Eleotris macrolepidotus, 795
Eleotris muralis, 794
Eleotris ophiocephalus, 795

Eleotris porocephalus, 795
Eleotris tæmifrons, 795
Eleotris viridis, 795
ellioti, Apogon, 784
ellioti, Carcharias, 809, 810
ellioti, Eleotris, 794
elongata, Olyra, 806
emarginatus, Pomacentrus, 801
endekatonia, Apogon, 784
Epinephelus polleni, 781
Epinephelus retouti, 780
Equula dentex, 790
Eupomacentrus lividus, 801
Exocœtus altipinnis, 807
Exocœtus katopron, 807

fasciatus, Apogon, 784
fasciatus, Chætodon, 786
fasciatus, Cirrhites, 788
fasciatus, Cirrhitichthys, 788
fasciatus, Corythoichthys, 808
fasciatus, Mugil, 800
fasciatus, Syngnathus, 808
fasciatus, Tetragonoptrus, 786
fax, Priacanthus, 784
ferdau, Caranx, 789
ferruginosus, Geniates, 804
Fierasfer, 805
Fierasfer homei, 805
filamentosus, Dentex, 786
flavimarginatus, Serranus, 782
flavo-cœruleus, Caranx, 789
flavus, Chætodon, 786
fluviatilis, Duxordia, 805
fluviatilis, Leiocassis, 805
fulungee, Cirrhina, 807
fulviflamma, Lutjanus, 783
furcatus, Aphareus, 782
furcatus, Labrus, 782
fuscus, Pimelepterus, 788
fuscus, Pomacentrus, 801
fuscus, Pseudochromis, 791
fuscus, Salarias, 797
fuscus, Serranus, 780
fuscus, Xyster, 788

gahm, Acanthurus, 789
Galaxias, 806
Galaxias indicus, 806
Galaxidæ, 806
garretti, Mesoprion, 783
Gazza argentaria, 790
Gazza tapeinosoma, 790
Geniates ferruginosus, 804
Genyoroge macolor, 783
Genyoroge nigra, 783
Genyoroge notata, 783
geoffroyi, Serranus, 780
Gerres altispinis, 786
Gerres setifer, 786
gibbosus, Serranus, 776
Ginglymostoma concolor, 811
Ginglymostoma rüppellii, 811
Glyphidodon alborinetus, 801
Glyphidodon anterius, 801
Glyphidodon ater, 801
Glyphidodon cingulus, 801
Glyphidodon cyaneus, 801
Glyphidodon dispar, 801
Glyphidodon henimelas, 801
Glyphidodon melas, 801
Glyphidodon modestus, 801
Glyphidodontops antjerius, 801
Glyphidodon unimaculatus, 801
Glyphidodon zonatus, 801
Glyphisodon leucopoma, 801
Glyphisodon phalosoma, 801
Glyphisodon xanthozona, 801

Gnathocentrum cantrogna-
thum, 786
Gobio angrioides, 807
Gobiodon quinque-strigatus,
794
Gobiodon rivulatus, 794
Gobius acutipinnis, 793
Gobius chlorostigma, 793
Gobius gymnocephalus, 792
Gobius histrio, 794
Gobius littoreus, 793
Gobius microlepis, 793
Gobius pleurostigma, 793
Gobius rivulatus, 794
Gobius sadanundio, 793
Gobius thurstoni, 793
Gobius viridipunctatus, 793
grammicus, Serranus, 780
Grammistes compressus, 783
Grammistes punctatus, 782
grandoculis, Chrysophrys, 787
grandoculis, Monotaxis, 787
grandoculis, Sciæna, 787
grandoculis, Sphærodon, 787
griseum, Diagramma, 785
guamensis, Psenes, 790
guamensis, Scorpæna, 788
guamensis, Scorpænopsis, 788
güntheri, Lunna, 810
guttatissimus, Chætodon, 786
guttatus, Serranus, 782
gymnocephalus, Gobius, 792

hæmatopterus, Syngnathus, 808
haffara, Chrysophrys, 788
haffara, Sparus, 788
hajam, Monacanthus, 808
halei, Cristiceps, 799
halei, Coris, 808
halei, Peristethus, 791
Halichœres javanicus, 803
hamrahr, Anthias, 783
hamrahr, Priacanthus, 783
hamrahr, Sciæna, 783
hasta, Sparus, 788
Heliastes reticulatus, 800
helleri, Monacanthus, 808
Hemerocœtes, 793
hemigymnostelhus, Cama-
goides, 789
hemisticta, Trigla, 791
hemistictus, Serranus, 782
henimelas, Glyphidodon, 801
heterodon, Sphærodon, 787
hippos, Caranx, 789
histrio, Gobius, 794
hoedtii, Eleotris, 794
hoedtii, Ophiocara, 795
Holocentrus Bengalensis, 783
Holocentrus malabaricus, 780
holocentrus, Priacanthus, 784
Holocentrus platyrhinum, 788
Holocentrus quinquelinearis,
783
Holocentrus quinquelinealus,
783
Holocentrus sammara, 788
holomelas, Salarias, 797
homei, Fierasfer, 805
homei, Oxybeles, 805
Hoplonotus, 791

immaculatus, Mesoprion, 783
impudicus, Caranx, 789
indica, Cepola, 790
indicus, Acanthoclinus, 793
indicus, Galaxias, 806
intermedius, Amphiprion, 800
irregularis, Scaphiodon, 807

janesaba, Scomber, 790
japonicus, Synagris, 786
javanicus, Halichœres, 803
javanicus, Julis, 803
javanicus, Platylossus, 803
javanicus, Psenes, 790
jayakari, Caranx, 789
jayakari, Diagramma, 785
jerdoni, Brotula, 804
Julis javanicus, 803
Julis metager, 802

Karun natacoli, 792
kasmira, Lutjanus, 783
katopron, Exocœtus, 807
kutunho, Pomacentrus, 801
kennedyi, Trachynotus, 790
klunzingeri, Mugil, 800
kurra, Caranx, 789

Labrus bilunulatus, 802
Labrus furcatus, 782
Labrus latovittatus, 787
Labrus spilonotus, 802
Lamna güntheri, 810
Laman spallanzanii, 810
Lates calcarifer, 779
latidens, Lethrinus, 787
latidens, Sphærodon, 787
latifasciatus, Serranus, 780, 781
latovittatus, Labrus, 787
latovittatus, Malacanthus, 787
latovittatus, Tæniunotus, 787
Leiocassis fluviatilis, 805
leonina, Scorpæna, 788
leonina, Scorpænopsis, 788
leopardus, Bleunius, 796
leopardus, Serranus, 782
Leptocardii, 812
Leptosynanccia, 788
lessonii, Diagramma, 785
lessonii, Nemophis, 799
Lethrinus acutus, 787
Lethrinus latidens, 787
Lethrinus miniatus, 787
Lethrinus olivaceus, 787
Lethrinus rostratus, 787
Lethrinus waigiensis, 787
leucopoma, Glyphidodon, 801
Leucopsarion Petersii, 800
Lichia toloparah, 789
lineata, Perca, 785
lineato-oculatus, Eleotris, 794
lineatus, Cheilodipterus, 785
lineatus, Salarias, 796
lineolatus, Acronurus, 789
lineolatus, Apogon, 785
littoreus, Gobius, 793
lividus, Chætodon, 801
lividus, Eupomacentrus, 801
lividus, Pomacentrus, 801
longicauda, Olyra, 806
loo, Scomber, 790
louti, Variola, 782
lunula, Chætodon, 786
Lutjanus argentimaculatus, 783
Lutjanus Bengalensis, 783
Lutjanus dentex, 782
Lutjanus fulviflamma, 783
Lutjanus kasmira, 783
Lutjanus macolor, 783
Lutjanus notatus, 783
Lutjanus nigra, 783
Lutjanus russellii, 783
Lutjanus quinquelineatus, 788
lysan, Chorinemus, 789

macolor, Diacope, 783
macolor, Genyoroge, 783

macolor, Lutjanus, 783
macolor, Mesoprion, 783
Macolor typus, 788
macracanthus, Priacanthus, 784
macrocephalus, Eleotris, 795
macrochilus, Mugil, 800
macrodon, Cheilodipterus, 785
macrodon, Paramia, 785
macrogenis, Serranus, 780
macrolepidota, Eleotris, 794
macrolepidotus, Eleotris, 795
macropterus, Apogon, 785
maculatus, Diodon, 809
madagascarensis, Xiphogadus, 799
madraspatensis, Priacanthichthys, 781
malabarica, Pempheris, 788
malabaricus, Caranx, 789
malabaricus, Holocentrus, 780
malabaricus, Serranus, 780
maldat, Cossyphus, 802
Malacanthidæ, 786
Malacanthus, 787
Malacanthus tæniatus, 787
Malacanthus latovittatus, 787
Malacocanthus, coccinicauda, 791
mangula, Pempheris, 786
marginatum, Tetradrachmum, 800
marginatus, Caranx, 789
marmoratus, Salarias, 798
mata, Acanthurus, 789
matoides, Acanthurus, 789
mauritiana, Chorinemus, 780
maximus, Apogon, 784
melampygus, Caranx, 789
melanurus, Acanthurus, 789
melas, Glyphidodon, 801
melas, Paraglyphidodon, 801
Melichthys ringens, 808
Mesoprion anrolineatus, 783
Mesoprion cæruleolineata, 783
Mesoprion deutex, 782
Mesoprion ehrenbergii, 783
Mesoprion garretti, 788
Mesoprion immaculatus, 788
Mesoprion macolor, 783
Mesoprion Russellii, 783
metager, Platyglossus, 802
microlepidotus, Scomber, 790
microlepis, Gobius, 793
microlepis, Oxyurichthys, 793
micrurus, Cossyphus, 802
miliaris, Chætodon, 796
miliaris, Tetragonoptrus, 786
miniatus, Sparus, 787
minutus, Sebastes, 788
moadetta, Chorinemus, 789
modestus, Glyphidodon, 801
mokarran, Sphyrna, 809
mokarran, Zygæna, 809
mola, Orthagoriscus, 809
molluccensis, Scomber, 790
molucca, Pempheris, 788
Monacanthus hajam, 808
Monacanthus helleri, 808
Monacanthus tormentosus, 808
Monacanthus trichurus, 808
Monotaxis grandoculis, 787
morrhua, Serranus, 780, 781
Mugil carinatus, 800
Mugil cirrhostomus, 800
Mugil crenilabris, 800
Mugil fasciatus, 800
Mugil klunzingeri, 800
Mugil macrochilus, 800
Mugil œur, 800

Mugil planiceps, 800
Mugil rüppellii, 800
Mugil tade, 800
multibarbata, Brotula, 804
multidens, Anthias, 782
multipunctatus, Serranus, 780
muralis, Eleotris, 794
murdjan, Myripristis, 788
murrayi, Carcharias, 809, 810
mutiens, Trichiurus, 788
myops, Ambassis, 784
Myripristis adusta, 768
Myripristis botche, 788
Myripristis murdjan, 788
Myxus superficialis, 800

nageb, Pristipoma, 785
Nebrius concolor, 811
neilli, Salarias, 797
Nemophis lessonii, 799
Nga yanga ap'lyoo, 801
niger, Balistes, 808
niger, Dascyllus, 801
niger, Proamblys, 783
niger, Salarias, 797
nigra, Diacope, 783
nigra, Genyoroge, 783
nigra, Lutjanus, 783
nigra, Sciæna, 783
nigrescens, Caranx, 789
nigricans, Acanthurus, 789
nigricans, Sparus, 801
notata, Genyoroge, 783
notatus, Ambassis, 784
notatus, Lutjanus, 783
notatus, Synagris, 786
novemcinctus, Serranus, 782
novem-maculatus, Diodon, 809
novem-maculatus, Paradiodon, 809
nuchalis, Pomacentrus, 801
nudus, Chætodon, 786

ocellata, Parachætodon, 786
ocellatus, Chætodon, 786
Odontaspis, 810
Odontaspis tricuspidatus, 810
œur, Mugil, 800
oligacanthus, Chætodon, 786
olivaceus, Lethrinus, 787
Olyra elongata, 806
Olyra longicauda, 806
oortii, Salarias, 798
Ophiocara aporos, 795
Ophiocara hoedtii, 795
Ophiocara ophiocephala, 795
Ophiocara tolsoni, 795
ophiocephala, Ophiocara, 795
ophiocephalus, Eleotris, 795
orientalis, Chorinemus, 789
orientalis, Corystion, 792
orientalis, Dactyloptorus, 792
Orthagoriscus, 809
Orthagoriscus mola, 809
Orthagoriscus oxyuropterus, 809
ovatus, Trachynotus, 790
Oxybeles, 805
Oxybeles brandesii, 805
Oxybeles homei, 805
oxycephala, Scorpænopsis, 788
Oxyurichthys microlepsis, 793
oxyuropterus, Orthagoriscus, 809

pacificus, Regalecus, 800
pæcilopterus, Apogon, 785
Pagrus ruber, 787
Pagrus spinifer, 787
pabijius, Cheilodipterus, 791

panijius, Sillago, 791
pantherinus, Serranus, 780
Parachætodon ocellatus, 786
Paradiodon novem-maculatus, 809
Paradiodon quadri-maculatus, 809
Paraglyphidodon melas, 801
Paramia macrodon, 785
parapistes, Caranx, 789
pectinatus, Pristis, 811
Pempheris malabarica, 788
Pempheris mangula, 788
Pempheris molucca, 788
Pempheris rhomboideus, 788
Pempheris Russellii, 788
Perca arabica, 785
Perca areolata, 780
Perca lineata, 785
Percis cylindrica, 790
Peristethus, 791
Peristethus halei, 791
petersii, Leucopsarion, 806
petersi, Petroscirtes, 796
Petroscirtes petersi, 796
Petroscirtes striatus, 796
Petroscirtes variabilis, 796
phaiosoma, Glyphidodon, 801
phaiosoma, Salarias, 797
plotos, Akysis, 806
pictus, Balistes, 808
Pimelepterus fuscus, 788
Pimelepterus waigiensis, 788
pinnulata, Scriola, 789
pinnulatus, Elagatis, 789
planiceps, Mugil, 800
Platyglossus juvenicus, 803
Platyglossus metager, 802
Platyglossus roseus, 803
platyrhinum, Holocentrus, 788
Plectorhynchus cuvieri, 785
Plectorhynchus schæ, 785
pleurostigma, Gobius, 793
pneumatophorus minor, Scomber, 790
polleni, Epinephelus, 781
polleni, Serranus, 781
polota, Datnioides, 786
polylepis, Sebastopsis, 788
polyophthalmus, Trichonotus, 796
Pomacentrus albofasciatus, 801
Pomacentrus bilineatus, 801
Pomacentrus chrysurus, 801
Pomacentrus emarginatus, 801
Pomacentrus katunbu, 801
Pomacentrus lividus, 801
Pomacentrus nuchalis, 801
Pomacentrus prosopotænia, 801
Pomacentrus punctatus, 801
Pomacentrus simsiang, 801
Pomacentrus tæniometopon, 801
Pomacentrus tæniops, 801
Pomacentrus trilineatus, 801
Pomacentrus trimaculatus, 801
Pomacentrus tripunctatus, 801
Pomacentrus unifasciatus, 801
Pomacentrus vanicolensis, 801
Porobronchus, 805
porocephalus, Eleotris, 795
præopercularis, Serranus, 780
Priacanthichthys madraspatensis, 781
Priacanthus blochii, 783
Priacanthus fax, 784
Priacanthus hamruhr, 783
Priacanthus holocentrum, 784
Priacanthus macracanthus, 784
Priacanthus schmittii, 784

Priacanthus tayenus, 784
pristipoma, Aprion, 782
pristipoma, Centopristis, 782
pristipoma, Chætopterus, 782
pristipoma, Dentex, 782
Pristipoma nageb, 785
Pristipoma stridens, 785
Pristipomoides typus, 782
Pristis pectinatus, 811
Pristotis fuscus, 801
Proamblys niger, 783
Prochilus bifasciatus, 800
Prochilus sebæ, 800
prosopotænia, Pomacentrus, 801
Psenes guamuensis, 790
Psenes javanicus, 790
Pseudochromis adustus, 791
Pseudochromis fuscus, 791
Pseudochromis xanthochir, 791
Pseudolates cavifrons, 779
Pseudosynanceia, 788
Pseudoscarus dussumieri, 804
Pseudoscarus bataviensis, 803
Pterois cincta, 788
Pterois radiata, 788
punctatus, Grammistes, 782
punctatus, Pomacentrus, 801

quadrifasciatus, Chætodon, 786
quadrifasciatus, Datnioides, 786
quadri-maculatus, Diodon, 809
quadri-maculatus, Paradiodon, 809
quinquelinearis, Holocentrus, 783
quinquelineatus, Holocentrus, 783
quinquelineatus, Lutjanus, 783
quinque-strigatus, Gobiodon, 794

radiata, Pterois, 788
ranga, Ambassis, 784
Raja rhinobatus, 812
Ragalecus pacificus, 800
Regalecus russellii, 800
retourti, Epinephelus, 780
reticulatus, Heliastes, 800
Rhinobatus annulatus, 811
Rhinobatus columnæ, 811
rhinobatus, Raja, 812
Rhinodontidæ, 811
Rhinodon typicus, 811
rhomboideus, Pempheris, 788
ringens, Melichthys, 808
rivulatus, Gobiodon, 794
rivulatus, Gobius, 794
robustus, Xiphochilus, 802
roseus, Platyglossus, 803
rostratus, Lethrinus, 787
ruber, Pagrus, 787
rubropunctata, Scorpæna, 788
rüppellii, Caranx, 789
rüppellii, Ginglymostoma, 811
rüppellii, Mugil, 800
russellii, Decapterus, 789
russellii, Lutjanus, 783
russellii, Mesoprion, 783
russellii, Pempheris, 788
russellii, Regalecus, 800
russellii, Trachynotus, 790
rutilans, Aphareus, 782

sadanundio, Gobius, 793
Salarias alboguttatus, 798
Salarias arenatus, 798
Salarias bicolor, 798
Salarias brevis, 796
Salarias caudolineatus, 798

Salarias cruentipinnis, 797
Salarias fuscus, 797
Salarias holomelas, 797
Salarias lineatus, 796
Salarias marmoratus, 796
Salarias neilli, 797
Salarias niger, 797
Salarias oortii, 796
Salarias phaiosoma, 797
Salarias sindensis, 797
Salarias steindachneri, 796
sancti-petri, Chorinemus, 780
savala, Trichiurus, 788
Scarus bataviensis, 808
Scarus dussumieri, 804
Scaphiodon irregularis, 807
schmittii, Priacanthus, 784
Sciaena grandoculis, 787
Sciaena hamruhr, 788
Sciaena nigra, 783
scie, Squalus, 811
Scomber janesaba, 790
Scomber loo, 790
Scomber microlepidotus, 790
Scomber moluccensis, 790
Scomber pneumatophorus
    minor, 790
Scorpaena chiliopristis, 788
Scorpaena guamensis, 788
Scorpaena leonina, 788
Scorpaena rubropunctata, 788
Scorpaenopsis guamensis, 788
Scorpaenopsis leonina, 788
Scorpaenopsis oxycephala, 788
Sebastes minutus, 788
Sebastopsis polylepis, 788
sebm, Amphiprion, 800
sebm, Diagramma, 785
sebm, Plectorhynchus, 785
sebm, Prochilus, 800
Seriola pinnulata, 789
Seriolichthys bipinnulatus, 789
Serranus aculirostris, 780
Serranus altivelis, 780
Serranus angularis, 780
Serranus areolatus, 780
Serranus diacanthus, 780
Serranus flavimarginatus, 782
Serranus fuscus, 780
Serranus geoffroyi, 780
Serranus gibbosus, 779
Serranus grammicus, 780
Serranus guttatus, 782
Serranus hemistictus, 782
Serranus latifasciatus, 780, 781
Serranus leopardus, 782
Serranus macrogenis, 780
Serranus malabaricus, 780

Serranus morrhua, 780, 781
Serranus multipunctatus, 780
Serranus novemcinctus, 782
Serranus pantherinus, 780
Serranus polleni, 781
Serranus praeopercularis, 780
Serranus sexmaculatus, 782
Serranus striolatus, 779
Serranus summana, 780
Serranus tinca, 780
Serranus tumilabris, 780
Serranus undulosus, 780
Serranus wandersi, 780
Serranus zanana, 782
setifer, Gerres, 780
setifer, Xiphasia, 790
setigerus, Trichonotus, 795
sex-maculatus, Diodon, 809
sexmaculatus, Serranus, 782
Sillago domina, 791
Sillago panijius, 791
simsiang, Pomacentrus, 801
sindensis, Salarias, 797
sinuata, Umbrina, 788
spallanzanii, Lamna, 810
Sparus cuvieri, 788
Sparus datnia, 788
Sparus haffara, 788
Sparus hasta, 788
Sparus mangula-kutti, 788
Sparus miniatus, 787
Sparus nigricans, 801
speciosus, Caranx, 789
Sphaerodon heterodon, 787
Sphaerodon latidens, 787
Sphyrna mokarran, 809
spilonotus, Labrus, 802
spinifer, Pagrus, 787
spinosissimus, Diodon, 809
Squalus scie, 811
Squalus vulpes, 810
steindachneri, Blennius, 796
steindachneri, Salarias, 796
stellatus, Caranx, 789
striata Umbrina, 788
striatus, Acanthurus, 789
striatus, Petroscirtes, 796
stridens, Pristipoma, 785
strigosus, Acanthurus, 789
striolatus, Serranus, 779
summana, Serranus, 780
summara, Holocentrus, 788
superficialis, Myxus, 800
Synagris japonicus, 786
Synagris notatus, 786
Synagris taeniopterus, 786
Syngnathus conspicillatus, 808

Syngnathus fasciatus, 808
Syngnathus haemalopterus, 808

tachete, Diodon, 800
tade, Mugil, 800
taeniatus, Malacanthus, 787
taeniometopon, Pomacentrus,
    801
Taenianotus latovittatus, 787
taeniope, Pomacentrus, 801
taeniopterus, Dentex, 786
taeniopterus, Synagris, 786
talamparoides, Carangoides, 780
tapeinosoma, Gazza, 790
tau nigrum, Chaetodon, 786
tayenus, Priacanthus, 784
tennentii, Acanthurus, 788
Tetradrachmum marginatum,
    800
Tetradrachmum trimaculatum,
    801
Tetragonoptrus fasciatus, 786
Tetragonoptrus miliaris, 786
Therapon, 785
thurstoni, Apogon, 784
tickelli, Apogon, 785
tinca, Serranus, 780
tol, Chorinemus, 789
toloo, Chorinemus, 789
tolsoni, Ophiocara, 795
tormentosus, Dalistes, 808
tormentosus, Monacanthus, 808
thurstoni, Gobius, 793
Trachynotus coppingeri, 790
Trachynotus kennedyi, 790
Trachynotus russellii, 790
Trachynotus ovatus, 790
Trichiurus auriga, 788
Trichiurus cristatus, 788
Trichiurus muticus, 788
Trichiurus savala, 788
trichurus, Monacanthus, 809
tricuspidatus, Carcharias, 810
tricuspidatus, Odontaspis, 810
Trichonotidae, 795
Trichonotus, 795
Trichonotus polyophthalmus,
    795
Trichonotus setigerus, 795
trifasciatum, Amphiprion, 800
trifasciatus, Chaetodon, 786
Trigla, 791
Trigla arabica, 791
Trigla hemisticta, 791
trilineatus, Pomacentrus, 801
trimaculatum, Tetradrachmum,
    801

trimaculatus, Dascyllus, 801
trimaculatus, Pomacentrus, 801
tripunctatus, Pomacentrus, 801
tristis, Acanthurus, 788
tumilabris, Serranus, 780
tumifrons, Eleotris, 795
typicus, Rhinodon, 811
typus, Macolor, 783
typus, Pristipomoides, 782

Umbrina strata, 788
Umbrina sinuata, 788
undulatus, Cheilinus, 802
undulatus, Crassilabrus, 802
undulosus, Serranus, 780
unicolor, Dascyllus, 801
unifasciatus, Pomacentrus, 800
unimaculatus, Glyphidodon, 801

vanicolensis, Pomacentrus, 801
variabilis, Petroscirtes, 796
Variola, louti, 782
venator, Caranx, 789
viridipunctatus, Gobius, 793
viridis, Eleotris, 795
vittatus, Chaetodon, 786
vittatus, Citharoedus, 786
vittatus, Coracinus, 800
vulpes, Alopecias, 810
vulpes, Alopias, 810
vulpes, Carcharias, 810
vulpes, Squalus, 810

waigiensis, Lethrinus, 787
waigiensis, Pimelepterus, 788
wandersi, Serranus, 780
wiebeli, Chaetodon, 786

xanthochir, Pseudochromis, 791
xanthopterus, Acanthurus, 789
xanthozona, Glyphidodon, 801
Xiphasia setifer, 790
Xiphochilus, 802
Xiphochilus robustus, 802
Xiphogadus madagascarensis,
    799
Xyster fuscus, 788

zanana, Serranus, 782
Zanclus, canescens, 786
Zanclus, centrognathus, 786
Zanclus cornutus, 786
Zeus argentarius, 790
zonatus, Glyphidodon, 801
Zygaena dissimilis, 810
Zygaena mokarran, 809

G. NORMAN AND SON, PRINTERS, HART STREET, COVENT GARDEN.

BY THE SAME AUTHOR.

THE LAND OF THE PERMAULS, OR COCHIN ITS PAST AND ITS PRESENT, 8vo. pages 577, 1863   Gantz Brothers, Madras, Rs. 10

THE FISHES OF MALABAR, 4to. pages 293, with 20 plates, coloured or plain, 1865
Bernard Quaritch, 15, Piccadilly, London.

REPORT ON THE FRESHWATER FISH AND FISHERIES OF INDIA AND BURMA, 8vo. pages 118 and 307 of Appendices, 1873   Government Press, Calcutta.

REPORT ON THE SEA FISH AND FISHERIES OF INDIA AND BURMA, 8vo. pages 86 and 332 of Appendices, 1873   Government Press, Calcutta.

THE FISHES OF INDIA, 4to. 2 volumes, pages 778, 200 plates, with upwards of a thousand lithographed figures, 1875-78, £12. 12s (a few copies remaining reduced to £6. 6s nett, at Messrs. Williams & Norgate)   B. Quaritch, 15, Piccadilly, London.

THE COMMERCIAL SEA FISHES OF GREAT BRITAIN, Prize Essay at the Great International Fisheries Exhibition, 8vo. pages 328, 1884, 5s   W. Clowes & Sons, 13, Charing Cross, London.

FISH CULTURE, Great International Fisheries Exhibition Series, 8vo. pages 105, with 4 plates
W. Clowes & Sons, 13, Charing Cross, London.

FOOD OF FISHES, Great International Fisheries Exhibition Series, 8vo. pages 36, 1883
W. Clowes & Sons, 13, Charing Cross, London.

CATALOGUE OF THE EXHIBITS IN THE INDIAN SECTION OF THE INTERNATIONAL FISHERIES EXHIBITION, 1883, 8vo. pages 197, 1883   Government of India, London.

THE FISHES OF GREAT BRITAIN AND IRELAND, a Natural History of such as are known to inhabit the Seas and Fresh-waters of the British Isles, their Economic Uses, Modes of Capture, &c., and an Introduction upon Fishes generally, imperial 8vo., cloth, 179 plates, 2 vols. 1880-1885 (pub. £5. 15s), £4. 18s   Williams & Norgate, 14, Henrietta Street, Covent Garden.

BRITISH AND IRISH SALMONIDÆ, imperial 8vo. cloth, 12 plates, some coloured, pages 208, 1887 (pub. £2. 2s), £1. 15s nett   Williams & Norgate, 14, Henrietta Street, Covent Garden.

www.ingramcontent.com/pod-product-compliance
Lightning Source LLC
Chambersburg PA
CBHW032033090426
42733CB00031B/1066